EINSTEIN AND RELIGION

MAX JAMMER

Einstein and Religion

PHYSICS AND THEOLOGY

PRINCETON UNIVERSITY PRESS

Published by Princeton University Press, 41 William Street,
Princeton, New Jersey 08540
In the United Kingdom: Princeton University Press,
Chichester, West Sussex

Library of Congress Cataloging-in-Publication Data

Jammer, Max.
Einstein and religion : physics and theology / Max Jammer.
p. cm.
Includes bibliographical references and index.
ISBN 0-691-00699-7 (cl : alk. paper)
1. Einstein, Albert, 1879–1955—Religion. 2. Religion
and science. I. Title.
QC16.E5J36 1999
215—dc21 99-24124

This book has been composed in Palatino

The paper used in this publication meets
the minimum requirements of
ANSI/NISO Z39.48-1992 (R1997)
(*Permanence of Paper*)

http://pup.princeton.edu

Printed in the United States of America

3 5 7 9 10 8 6 4 2

Contents

Acknowledgments

I WISH to acknowledge my indebtedness to the Institute for Advanced Study in Princeton, New Jersey, and to the Einstein Archive at the National and University Library in Jerusalem for having permitted me to examine and to use Einstein's still unpublished writings. I also express my gratitude to the Mosad Harav Kook in Jerusalem, the Union Theological Seminary in New York, and the National Academy of Sciences in Washington, D.C., which I consulted in the course of my work. It is a pleasure to express my appreciation to Professor John Stachel of the Department of Physics and Center for Einstein Studies, Boston University, and to Dr. Trevor Lipscombe, Physics Editor at Princeton University Press, for their encouragement to write this book. Finally, I wish to thank Teresa Carson for her careful editing of the manuscript, Lys Ann Shore, Ph.D., for her meticulous preparation of the index, and Jane Low, manager of the editorial production group at Princeton University Press, for the fruitful cooperation.

EINSTEIN AND RELIGION

Introduction

ALBERT EINSTEIN is generally regarded as the greatest theoretical physicist of the twentieth century or "possibly of all times."[1] In any case, modern physics bears his impact more than that of any other physicist. His contributions to atomic physics—among them, his study of the photoelectric effect, for which he was awarded the Nobel Prize, and his theory of relativity with its profound modifications of the notions of space, time, and gravitation—have fundamentally changed and deepened our physical and philosophical conception of the universe. Apart from his scientific ingenuity, his courageous struggle for human rights, social justice, and international peace has assured him a unique place in the history of our age.

It is therefore not surprising that a great number of biographies and historical studies have been written about Einstein's life and work. One of them, published in 1966, stated that "in recent years some four hundred books about Einstein and his work have appeared."[2] In fact, since 1991 alone, well over a dozen Einstein biographies have been published.[3]

[1] R. W. Clark, *Einstein—The Life and Times* (Avon, New York, 1971), p. 10.

[2] D. Brian, *Einstein—A Life* (Wiley, New York, 1966), p. IX.

[3] J. Bernstein, *Einstein* (Fontana Press, London, 1991, 1993); R. Schulmann and J. Renn, *Albert Einstein—Mileva Maric* (Princeton University Press, Princeton, N.J., 1992); R. B. Dilts, *Einstein* (Junfermann, Paderborn, 1992); J. Merleau-Ponty, *Einstein* (Flammarion, Paris, 1993, 1995); M. White and J. Gribben, *Einstein* (Simon and Schuster, New

Yet, none of these biographies, including the recent publications on previously concealed facts about Einstein's private life, gives an adequate account, if any, of an important facet of his life: his undogmatic and yet profound religiosity and his philosophy of religion. Even John Stachel's excellent piece of documentary research on Einstein's Jewish identity deals, as does Gerald E. Tauber's *Einstein on Zionism, Arabs and Palestine*, with Einstein's conception of Judaism primarily from the sociopolitical point of view.[4]

In some biographies, especially those written while Einstein was still alive, occasional references to his early religiosity can be found. But throughout Einstein's life, including his later years, religious sentiments and theological reflections played a role of much greater importance than any of his biographers seems to have realized.

To prove this contention and thus fill an important lacuna in the biographical literature about Einstein is one of the objectives of this monograph. Were it the only aim, the

---

York, 1993); B. Kirksberg, *Einstein: Humanismo y Judaismo* (Lumen, Buenos Aires, 1993); A. Fölsing, *Albert Einstein* (Suhrkamp, Frankfurt, 1993; Viking, New York, 1997); R. Highfield and P. Carter, *The Private Lives of Albert Einstein* (Faber and Faber, London, St. Martin's Press, New York, 1993) (German ed., Deutscher Taschenbuch Verlag, Munich, 1996); F. Balibar, *Einstein—La Joie de la Pensée* (Éditions Gallimard, Paris, 1993) (German ed., Ravensburger Buchverlag, Ravensburg, 1995); A. Pais, *Einstein Lived Here* (Oxford University Press, Oxford, 1994); A. Hermann, *Einstein—Der Weltweise und sein Jahrhundert* (Piper, Munich, 1994); G. Holton, *Einstein, History and Other Passions* (American Institute of Physics Press, New York, 1995); D. Brian, *Einstein—A Life* (Wiley, New York, 1996).

[4] J. Stachel, "Einstein's Jewish Identity" (draft prepared for the Symposium "Einstein in Context," Center for Einstein Studies and Physics Department, Boston University, Boston, 1989); G. E. Tauber, *Einstein on Zionism, Arabs and Palestine* (Tel-Aviv University, Tel-Aviv, 1979).

title of the book would have been *Einstein's Religion,* not *Einstein and Religion.*

This monograph intends to study not only how deeply religion affected Einstein and his work, but also conversely, how deeply Einstein's work, and in particular his theory of relativity, affected theological thought, a problem that has not yet been explored systematically. The investigation of such a possible interaction between modern physics and theology is, of course, apt to throw new light on the much discussed controversy about the relation between religion and science, an issue that was also of cardinal importance in Einstein's philosophy of religion.

The present monograph is a revised and considerably enlarged version of my booklet, *Einstein und die Religion,* published in 1995 by the Universitätsverlag Konstanz in Germany. The German contains a foreword by the noted physicist and cosmologist Jürgen Audretsch and an epilogue by the well-known physicist and philosopher Carl Friedrich von Weizsäcker. Like the present monograph, it contains three different, though interrelated, chapters. The first chapter describes Einstein's personal attitude toward religion from his early youth until his death and thus supplements the existing Einstein biographies. It is based on a lecture that I delivered on October 26, 1993, at Einstein's summer house in Caputh, a small town in the state of Brandenburg near Potsdam on the outskirts of Berlin. Einstein bought this block house with his life savings in 1929 and stayed there for the last three summers before he left Germany in 1932, when he fled the impending Nazi terror and never returned to his native country. The dramatic history of this simple wooden building, the only physical reminder of the great physicist's presence in Germany, has been vividly recounted by Michael Grüning.[5] A resident of Princeton, New Jersey, since 1933, Einstein became, in 1949, an honorary citizen of Caputh, which at that time belonged to the German Democratic Republic (DDR).[6] After the reunification of Germany, the "Einstein House" became a center for cultural activities.

In October 1993, Caputh celebrated the 675th anniversary of its foundation. In order to commemorate Caputh's most famous former citizen on this occasion, the minister of science in the state of Brandenburg, together with Caputh's Bürgermeister Dr. Grütte, invited me, because of my personal acquaintance with Einstein, to deliver a lecture on Einstein. Because the lecture should convey something new and also be comprehensible to nonphysicists, I decided—in consultation with Caputh's Pastor, Dr. H. Heilmann, advocate Ed Dellian, and Dr. Gary Smith, the director of the Einstein Forum—that "Einstein's religion" would be an appropriate topic. I accepted the invitation although I was at first not quite sure whether this topic would be sufficiently thought-provoking for such an occasion. I regretted, of course, that I had never raised this subject in any of my conversations with Einstein. When I consulted the Einstein Archive at the Hebrew National and University Library in Jerusalem and other sources, I soon realized that religion played an important role in Einstein's emotional and intellectual life and that the Swiss novelist and playwright Friedrich Dürrenmatt was not all wrong when

[5] M. Grüning, *Ein Haus für Einstein* (Verlag der Nation Berlin, Berlin, 1990).

[6] D. Charles, "Who owns Einstein's summerhouse?" *New Scientist* 154, no. 2078 (1997): 5.

he said, "Einstein used to speak so often of God that I tend to believe he has been a disguised theologian."[7]

In fact, the audience response at the Einstein House and the coverage of the lecture in the press showed that the topic touched upon issues whose relevance exceeded by far the confines of a specific biographical report and raised problems of general interest to every critically thinking person.[8] I therefore decided to discuss this theme in greater detail and wrote *Einstein und die Religion*. Encouraged by the press reviews of that book, I thought it appropriate to write a considerably enlarged edition in English.[9]

As mentioned above, the first chapter deals with Einstein's personal attitude toward religion from his school days until his death in 1955. The second chapter discusses what Einstein has written in his essays and in his correspondence about the nature of religion and its role in human society. The third and final chapter studies the influence, if any, of his scientific work on theological thought.

It cannot be emphasized too strongly that this study has no missionary intention whatever and does not attempt to convert the reader to Einstein's concept of religion. Nor does it intend even to defend his position or his philosophy of religion. It rather tries to discuss the issue *sine ira et studio*, in a completely unbiased way and in a historical

[7] "Einstein pflegte so oft von Gott zu reden, dass ich beinahe vermute, er sei ein verkappter Theologe gewesen." F. Dürrenmatt, *Albert Einstein* (Diogenes Verlag, Zürich, 1979), p. 12.

[8] See, e.g., the review of the lecture in the *Berliner Morgenpost* of 28 October 1993, p. 37, or in the periodical *Der Havelbote* of 17 November 1993, p. 1.

[9] See, e.g., the full-page article in the *Frankfurter Allgemeine* of 20 September 1995 (no. 219/38D).

and philosophical perspective without judgment of any kind. Were he alive today, Einstein would most certainly endorse such an approach. For he would never agree to proselytize people to his own religious conviction.

Anyone, irrespective of religious persuasion, can study Einstein's philosophy of religion without identifying with it or adopting it. The following historical facts clearly illustrate this point. In 1923, on his return journey from Japan, Einstein visited the Holy Land and received, while in Tel Aviv, a letter from Abraham J. Kook, the chief rabbi of the country since 1921 and the acknowledged leader of Jewish orthodoxy.[10] Apart from the traditional welcome wishes, it contained an invitation to meet the rabbi in Jerusalem. Although Einstein accepted, neither Einstein nor Kook seems to have mentioned this meeting in their reminiscences. We know from the minutes, taken by the rabbi's secretary, S. B. Schulman, that it was a very friendly and mutually respectful exchange of thoughts and opinions mainly on the interpretation of religious writings such as the esoteric doctrine of the Cabala.[11] This shows that even a strictly observant orthodox Jew should not place Einstein's religious philosophy under taboo. That a devout Christian also may respect Einstein's philosophy of religion is plainly exemplified by the Very Reverend Thomas Forsyth Torrance who served as a chaplain, as Moderator of the General Assembly of the Church of Scotland, and as Professor of Christian Dogmatics. According to Torrance, who holds doctorates in Divinity, Literature, and Science, Christian

[10] Rabbi Kook to Einstein, 4 February 1923. *Igroth Harav Kook* (in Hebrew) (Kook Institute, Jerusalem, 1984), p. 150.

[11] M. J. Zuriel, ed., *Ozroth Harav Kook* (Jeshivat Scha'alwim, Sha'alwim, 1988), p. 26.

theology is a positive science, based on a profound convergence of religious and scientific thought, a thesis that he shared with Einstein. Because this issue is included in our discussion of the implications of Einstein's work for theology, suffice it to point out that, on Torrance's initiative and under his general editorship, the Scottish Academic Press in Edinburgh published in its series "Theology and Science at the Frontiers of Knowledge" works like Ian Paul's *Science and Theology in Einstein's Perspective* (1982), Thomas F. Torrance's *Reality and Scientific Theology* (1985), and Ralph G. Mitchell's *Einstein and Christ: A New Approach to the Defence of the Christian Religion* (1987), all of which proclaim the thesis of a confluence of Einstein's ideas with orthodox theological thought.

Turning finally to Islam, the third of the great monotheistic religions, we can see that, in spite of the general reluctance of Muslim intellectuals to be drawn into the debate about science versus religion, Einstein was held in high honor by philosophers and "ulama" (religious teachers). Thus, for example, one of the earliest publications in Arabic on Einstein, Subhi Raghib's booklet *Unknown Facts about the Universe*, opens with a poem in which Einstein is called a "wise man" and compares him with an "Iman," a priest in a Muslim mosque.[12] In his explanation of Einstein's theory, Raghib quotes verses from the Koran in support of the theory and adds that Einstein revealed only a small part of what God has revealed in the Koran. Another example is the Egyptian philosopher Mahmoud Abbas al-Aqqad, who in his article, "The Understood Einstein," refers to difficulties in understanding Einstein's

[12] S. Raghib, *Haqaiq Majhoulah an al-Kawn* (Homes, Syria, 1927).

theory of relativity, in particular, the notion of four-dimensional space-time. Notions like a fourth dimension or the bounds of the universe, he wrote, "can only be described by a mathematician's hypothesis or by religious faith. There is really no difference between the alternatives, because mathematical assumptions and religious faith are both based on surrender and acceptance."[13]

Much effort has been spent on making the text accessible not only to the professional scientist but also to the nonspecialist interested in the relation between religion and science. The first two chapters can be understood without any knowledge of mathematics or physics because they deal almost exclusively with historical, philosophical, or theological issues, but a full comprehension of the third chapter requires some familiarity with the foundations of modern physics. Because as Einstein once said, "Most of the fundamental ideas of science are essentially simple, and may, as a rule, be expressed in a language comprehensible to everyone,"[14] almost all technicalities have been explained in simple terms without sacrificing logical rigor. For the few cases in which this could not be done, explanatory comments and bibliographical references have been provided.

Because the presentation of such a subject is easily susceptible to personal bias, extensive use has been made of quotations from the sources so as to avoid, as far as possible, any misinterpretations. The extensive documentation of the text therefore fulfills two functions: It not only en-

[13] M. A.. Al-Aqqad, "Einstein al-Mafhoum," *Al-Muqtataf* 75 (1929): 16–22.

[14] A. Einstein and L. Infeld, *The Evolution of Physics* (Simon and Schuster, New York, 1938), p. 27.

ables the reader to supplement his knowledge on a particular issue but also to verify that the presentation has not been prejudiced by the author.

It is of vital importance to understand how the three chapters of the book are internally related. Chapter 2, which discusses Einstein's philosophy of religion, can be regarded to a large extent as a logical justification of his personal attitude toward religion as described in chapter 1. But chapter 3, which deals with the alleged theological implications of his scientific work, should by no means be conceived as a logical justification of his philosophy of religion, and not only because all these implications were published after his death. Even though they reflect the opinions of prominent theologians and scientists, the arguments in chapter 3 are of a highly controversial nature. These arguments certainly form an interesting chapter in the history of ideas and are intimately related to Einstein's philosophy of religion, but they are not an integral part of it. In fact, although Einstein epitomized his philosophy of religion by stating that "science without religion is lame; religion without science is blind," he never based his religion on logical inferences from his scientific work. It is possible that he would have rejected all of the arguments in chapter 3 if he were alive. One of the major objectives of this study is to clarify Einstein's conception of religion by explaining this apparent contradiction.

CHAPTER 1

Einstein's Religiosity and the Role of Religion in His Private Life

In his autobiography, Einstein wrote that "the essential in the being of a man of my type lies precisely in *what* he thinks and *how* he thinks, not in what he does or suffers."[1] Had we strictly complied with this statement, we would have had to restrict our discussion on Einstein's thought about religion and the arguments on which he based his religious belief. But because a religious credo is usually conditioned, partially at least, by the milieu in which one grows up, by the education one receives, and by the literature one has read, we shall begin with an account of these factors insofar as they are relevant to Einstein's religious outlook.

Official records and Jewish family registers reveal that, since at least 1750, Einstein's paternal and maternal ancestors had lived in southern Germany, mainly in Buchau, a small town not far from Ulm. Albert's great-grandfather was born there in 1759, his grandfather Abraham in 1808, and his father Hermann in 1847. The fact that Albert, born in Ulm on March 14, 1879, was, contrary to Jewish tradition, not given the name of his grandfather, shows that his parents were not dogmatic in matters of religion. Although they never renounced their Jewish heritage, they did not observe traditional rites or dietary laws and never attended religious service at the synagogue. Hermann Einstein regarded Jewish rituals as relics of an ancient superstition and "was proud that Jewish rites were not practiced in his home," as Albert's son-in-law Rudolf Kayser wrote in his biography of Einstein, which he published under the pseudonym Anton Reiser.[2]

[1] A. Einstein, "Autobiographical Notes," in *Albert Einstein: Philosopher-Scientist*, ed. P. A. Schilpp (Library of Living Philosophers, Evanston, Ill., 1949), p. 33.

[2] A. Reiser, *Albert Einstein—A Biographical Portrait* (A. and C. Boni, New York, 1930), p. 28.

In June 1880, Hermann Einstein with his wife Pauline, née Koch, and the infant Albert moved from Ulm to Munich, the capital of Bavaria. Five months later, Maja, Albert's only sibling, was born. When Albert, at age six, entered the Petersschule, a Catholic public primary school (Volksschule), he received religious instruction, which at that time was compulsory in Bavaria. Although his parents were not observant, they hired a distant relative, whose name is not known, to teach Albert the principles of Judaism, obviously to counterpoise the Catholic instruction at school. According to Maja's recollection, it was this relative who awakened in the young Albert a fervent religious sentiment.

> He heard about divine will and works pleasing to God, about a way of life pleasing to God—without these teachings having been integrated into a specific dogma. Nevertheless, he was so fervent in his religious feelings that, on his own, he observed religious prescriptions in every detail. For example, he ate no pork. This he did for reasons of conscience, not because his family has set such an example. He remained true to his self-chosen way of life for years. Later religious feeling gave way to philosophical thought, but absolutely strict loyalty to conscience remained a guiding principle.[3]

A somewhat different explanation of young Albert's religious enthusiasm has been given by Alexander Moszkow-

[3] Maja Winteler-Einstein, "Albert Einstein—Beitrag für sein Lebensbild," *The Collected Papers of Albert Einstein* (J. Stachel, ed.), vol. 1 (Princeton University Press, Princeton, N.J., 1987), pp. xlvii–lxvi; "Albert Einstein—A Biographical Sketch" (translated excerpts), English translations of *The Collected Papers of Albert Einstein*, vol. 1 (Princeton University Press, Princeton, N.J., 1987), pp. xv–xxii.

ski, who wrote the first biography of Einstein in 1920. Based on personal conversations with Einstein, Moszkowski declared,

> His father, who had a sunny, optimistic temperament, and was inclined toward a somewhat aimless existence, at this time moved the seat of the family from Ulm to Munich. They here lived in a modest house in an idyllic situation and surrounded by a garden. The pure joy of Nature entered into the heart of the boy, a feeling that is usually foreign to the youthful inhabitants of cities of dead stone. Nature whispered song to him, and at the coming of the spring-tide infused his being with joy, to which he resigned himself in happy contemplation. A religious undercurrent of feeling made itself manifest in him, and it was strengthened by the elementary stimulus of the scented air, of buds and bushes, to which was added the educational influence of home and school. This was not because ritualistic habits reigned in the family. But it so happened that he learned simultaneously the teachings of the Jewish as well as the Catholic Church; and he had extracted from them that which was common and conducive to a strengthening of faith, and not what conflicted.[4]

In contrast to Maja's report that the private tutor stimulated in Albert religious feelings, Moszkowski claimed that the beauty and splendor of nature opened the gate of the

[4] A. Moszkowski, *Einstein—Einblicke in seine Gedankenwelt, entwickelt aus Gesprächen mit Einstein* (Hoffmann and Campe, Hamburg, 1920); *Einstein the Searcher—His Work Explained from Dialogues with Einstein* (Methuen, London, 1921), p. 221.

"religious paradise," as Einstein once called this phase of his youth. Moszkowski pointed out that yet another factor played an important role in Albert's religious feeling, and that was music. Ever since he took violin lessons at age six, Einstein found music intimately related with religious sentiments.

> Signs of his love for music showed themselves very early. He thought out little songs in praise of God, and used to sing them to himself in the pious seclusion that he preserved even with respect to his parents. Music, Nature, and God became intermingled in him in a complex of feeling, a moral unity, the trace of which never vanished, although later the religious factor became extended to a general ethical outlook on the world. At first he clung to a faith free from all doubt, as had been infused into him by the private Jewish instruction at home and the Catholic instruction at school. He read the Bible without feeling the need of examining it critically; he accepted it as a simple moral teaching and found himself little inclined to confirm it by rational arguments as his reading extended very little beyond its circle.[5]

That "Music, Nature, and God became intermingled in him in a complex of feeling" may well serve as a leitmotiv in this study of Einstein's religiosity. His conception of the relation between Nature and God will engage our attention throughout the discussions. The following episode illustrates how music and God were related in Einstein's mind.

[5] Moszkowski, *Einstein the Searcher*, p. 222.

On April 12, 1930, the Berlin Philharmonic Orchestra, conducted by Bruno Walter, gave a concert in Berlin. The program was Bach, Beethoven, and Brahms, and the soloist was Yehudi Menuhin. At the end of the recital, the audience burst into wild applause, and Einstein rushed over to Menuhin, embraced him, and exclaimed, "Now I know there is a God in heaven!"[6]

Because Moszkowki's book is essentially a report on conversations with Einstein, Einstein's own account of his early religiosity should fully agree with Moszkowski's report. Surprisingly, this is not the case. In his 1949 autobiographical notes, Einstein wrote:

> when I was a fairly precocious young man, the nothingness of the hopes and strivings which chases most men restlessly through life came to my consciousness with considerable vitality. Moreover, I soon discovered the cruelty of that chase, which in those years was more carefully covered up by hypocrisy and glittering words than is the case today. By the mere existence of his stomach, everyone was condemned to participate in that chase. Moreover, it was possible to satisfy the stomach by such participation, but not man insofar as he is a thinking and feeling being. As the first way out, there was religion, which is implanted into every child by way of the traditional education machine. Thus I came—despite the fact that I was the son of entirely irreligious (Jewish) parents—to a deep religiosity.[7]

[6] See, e.g., D. Brian, *Einstein—A Life*, p. 193.

[7] A. Einstein, "Autobiographical Notes," in *Albert Einstein: Philosopher-Scientist*, p. 3.

According to Einstein's recollection, the root of his religiosity, as we see, was neither a love of nature nor music; it was rather his realization of the vanity of human rivalry in the struggle for existence with its concomitant feeling of depression and desperation from which religion seemed to offer a relief. Such an attitude toward life can hardly have been entertained by a young boy, however. It seems, therefore, that Einstein's account is rather a projection of ideas pertaining to his mature age into his youth.

Historical surveys of Munich's educational system and other sources provide some information about the curriculum of Einstein's religious instruction at the Petersschule as well as at the Luitpold Gymnasium, the secondary school in which he enrolled in the beginning of 1888.[8] At the Catholic primary school, he was taught, at age seven, parts of the Small Catechism (Catechismus Romanus) and biblical tales of the New Testament; at age eight, sections of the Large Catechism and biblical stories of the Old Testament; and at age nine years, other parts of the Old Testament and the sacraments, baptism, and the Lord's Supper. As the only Jew in his class, Albert seemed never to have felt uncomfortable—with the possible exception of one incident. In one of these lessons, the teacher, a Catholic priest, held up a big nail and reportedly said that "these were the nails with which Christ was crucified by the Jews." According to the biographers, Rudolf Kayser and Carl Seelig, whose report is based mainly on correspondence with Einstein, the teacher intended to stir up hatred against the Jews, and all eyes in

[8] J. Gebele, *Hundert Jahre der Münchener Volksschule* (in German) (C. Gerber, Munich, 1903). Cf. also, Appendix A (Munich Volksschule, Curriculum) and Appendix B (Luitpold Gymnasium, Curriculum) in *The Collected Papers of Albert Einstein*, vol. 1, pp. 341–355.

the class turned to Albert who felt very embarrassed.[9] "For the first time Albert experienced the frightful venom of anti-Semitism," wrote Kayser (Reiser).

A somewhat different account of this episode can be found in Philipp Frank's biography of Einstein. According to Frank, the teacher said only, "The nails with which Christ was nailed to the cross looked like this," pointing to the nail he had brought. And Frank explicitly continued:

> But he did not add, as sometimes happens, that the Crucifixion was the work of the Jews. Nor did the idea enter the minds of the students that because of this they must change their relations with their classmate Albert. Nevertheless Einstein found this kind of teaching rather uncongenial, but only because it recalled the brutal act connected with it and because he sensed correctly that the vivid portrayal of brutality does not usually intensify any sentiments of antagonism to it but rather awakens latent sadistic tendencies.[10]

Frank's biography is known to be based largely on epistolary correspondence, whereas Kayser's account is based on personal conversations with Einstein. In his brief preface to Kayser's biography, Einstein declared, "I found the facts of the book duly accurate, and its characterization, throughout, as good as might be expected of one who is perforce himself, and who can no more be another than I

[9] A. Reiser, *Albert Einstein—A Biographical Portrait,* p. 30; C. Seelig, *Albert Einstein* (Europa Verlag, Zurich, 1960), p. 16.

[10] P. Frank, *Einstein—His Life and Times* (Knopf, New York, 1947), pp. 9–10.

can."[11] It is, of course, difficult today to find out which of the two versions is true. It is also difficult to assess how such an anti-Semitic incident, had it really happened, would have affected Albert's religious attitude toward Judaism.

In any case, Albert seemed to have liked these courses and on some occasions even helped his Catholic classmates when they failed to find the correct answer. Nor did he seem to have sensed any difference between what he learned about the Catholic religion at school and about the Jewish religion at home. He learned to respect sincere religious convictions of whatever denomination, an attitude he did not abandon in his later life when he rejected any affiliation with an institutional religious organization.

This attitude is evidenced in his replies to some questions raised by George Sylvester Viereck during a 1929 interview.

> "To what extent are you influenced by Christianity?"
>
> "As a child I received instruction both in the Bible and in the Talmud. I am a Jew, but I am enthralled by the luminous figure of the Nazarene."
>
> "Have you read Emil Ludwig's book on Jesus?"
>
> "Emil Ludwig's *Jesus* is shallow. Jesus is too colossal for the pen of phrasemongers, however artful. No man can dispose of Christianity with a *bon mot*!"
>
> "You accept the historical existence of Jesus?"
>
> "Unquestionably! No one can read the Gospels without feeling the actual presence of Jesus. His personality pulsates in every word. No myth is filled with such life."[12]

[11] A. Reiser, *Albert Einstein—A Biographical Portrait*, p. v.

[12] G. S. Viereck, "What Life Means to Einstein," *Saturday Evening*

The arrangement of religious instruction at the Luitpold Gymnasium differed from that at the Catholic Volksschule in several respects. As an interdenominational school, the Gymnasium offered special courses of religious instruction to its Jewish pupils. In contrast to the three weekly hours at the Petersschule, only two hours per week were devoted to religious studies, and these were given by external teachers especially ordained for this purpose by the Jewish community of the city. Einstein's first teacher was Herr Heinrich Friedmann. In his classes, which were shared by Einstein's Jewish classmates and the Jewish pupils of his next higher grade, Friedmann taught the Ten Commandments, biblical history, selected chapters of the Old Testament, the rituals of the Jewish holy days, and the rudiments of Hebrew grammar. From 1892 to 1895, the year Albert left Munich to join his parents in Italy without having completed his schooling, his teachers of religion were Dr. Joseph Perles, Eugene Meyer, and Dr. Cossmann Werner. They introduced him to the literature of the Psalms, and the history of the Talmud and of the Jews in Spain. Unfortunately, because these external teachers did not enjoy the same authority as their full-time colleagues at the Gymnasium, the attitude of their pupils toward their lessons seems to have been less serious that it should have been. Einstein referred to this in 1929 when he received fiftieth-birthday congratulations from his old teacher Heinrich Friedmann. Einstein declared: "I was deeply moved and delighted by your congratulations. How vividly do I remember those days of my youth in Munich and how deeply do I regret

Post, 26 October 1929; *Schlagschatten, Sechsundzwanzig Schicksalsfragen an Grosse der Zeit* (Vogt-Schild, Solothurn, 1930), p. 60; *Glimpses of the Great* (Macauley, New York, 1930), pp. 373–374.

not having been more diligent in studying the language and literature of our fathers. I read the Bible quite often, but the original text remains inaccessible for me. It certainly was not your fault; you have fought valiantly and energetically against laziness and all kinds of naughtiness."[13]

Einstein could have added that neither had it been Friedmann's fault nor the fault of any other of his teachers of religion that, at the age of twelve, just when he should have been preparing for the bar mitzvah, the Jewish confirmation, he suddenly became completely irreligious. Ironically, this conversion was, indirectly at least, the consequence of the only religious custom that his parents observed, namely to host a poor Jewish student for a weekly meal. The beneficiary was Max Talmud, a medical student from Poland, ten years older than Albert. In spite of their age difference, Albert and Talmud became intimate friends, and this friendship changed Albert's attitude toward religion. Because Talmud (or Talmey, as he called himself later when working as a general practitioner in New York) wrote a book on relativity in which he described his visits to the Einsteins in Munich, we have an authentic account of the influence he exerted on Albert.[14] He directed Albert's attention to Aaron Bernstein's popular

[13] "Ihre Gratulation hat mich gerührt and gefreut. Wie lebhaft sind mir die Münchener Jugendtage aus der Vergangenheit aufgestiegen und wie oft habe ich es bedauert, nicht fleissiger gewesen zu sein im Studium der Sprache und Literatur unserer Väter. Oft lese ich in der Bibel, aber der Urtext ist mir unzugänglich geblieben. Ihre Schuld ist es wahrlich nicht; Sie haben wacker und energisch gegen Faulheit und Allotria gekämpft." Einstein Archive, reel 30-403.

[14] M. Talmey, *The Relativity Theory Simplified* (Falcon Press, New York, 1932).

Naturwissenschaftliche Volksbücher (*Popular Books on Physical Science*), Ludwig Büchner's materialistic *Kraft und Stoff* (*Force and Matter*), Immanuel Kant's *Kritik der Reinen Vernunft* (*Critique of Pure Reason*) as well as to various books on geometry and other branches of mathematics. Einstein himself summed up the results of Talmey's influence:

> Through the reading of popular scientific books I soon reached the conviction that much in the stories of the Bible could not be true. The consequence was a positively fanatic [orgy of] freethinking coupled with the impression that youth is intentionally being deceived by the state through lies; it was a crushing impression. Suspicion against every kind of authority grew out of this experience, a skeptical attitude towards the convictions which were alive in any specific social environment—an attitude which has never again left me, even though later on, because of a better insight into the causal connections, it lost some of its original poignancy.[15]

An immediate consequence of this change of mind was the fact that Einstein refused to become bar mitzvahed.[16] Although this ceremonious act, introduced in the thirteenth century, is not a "halachist" (necessary) condition for membership in the Jewish community, even liberal Jews regard it as a precept that must be obeyed. By not complying with it, Einstein obviously intended to demonstrate his personal independence from the dictates of traditional authority. The

[15] A. Einstein, "Autobiographical Notes," in *Albert Einstein: Philosopher-Scientist*, p. 5.

[16] A. Pais, *Subtle is the Lord . . . The Science and the Life of Albert Einstein* (Oxford University Press, Oxford, 1982), p. 38.

nonperformance of his bar mitzvah would have caused serious political problems, at least on the part of the orthodoxy, had Einstein accepted David Ben-Gurion's offer in November 1952 to become the second president of the State of Israel after the death of Chaim Weizmann.

Interestingly, when he was living in Berlin, Einstein did own a pair of phylacteries (tephillin). Needless to say, Einstein never performed the ritual of putting them on as religious Jews used to do after becoming bar mitzvah. He kept them obviously only as an heirloom or memento of his ancestors. In May 1933, four months after Einstein had left Germany, his apartment on Haberlandstrasse 5 was raided by the Gestapo under the pretext of searching for anti-German propaganda literature, and these phylacteries and a prayer book, together with valuable pictures and cutlery, were looted.[17]

Einstein's indifference concerning religious affiliations is also shown by the fact that his first wife Mileva Maric, a fellow student at the Polytechnic in Zurich, belonged to the Greek Orthodox Church. Their marriage took place in Zurich in 1903 and was a civil ceremony without the presence of a rabbi or a priest. Both sets of parents had strongly opposed the marriage, mostly because of the difference in their religious backgrounds. After their two sons, Hans Albert and Eduard, were born, questions arose regarding their religious instruction and therefore their elementary school education.[18] Einstein reportedly said, "Anyway, I dislike very

[17] For details, see A. Hermann, *Einstein—Der Weltweise und sein Jahrhundert* (Piper, Munich, 1994), p. 410.

[18] The existence of their illegitimate daughter Lieserl, born in 1902 and apparently left with Mileva's relatives, became generally known only in 1987. In spite of careful research, no details about her fate are

much that my children should be taught something that is contrary to all scientific thinking."[19]

As far as we know, Einstein never attended religious service and never prayed in a synagogue or at any other place of worship. He visited such places only to participate in social events. The following examples illustrate this fact. On January 29, 1930, he participated at a Welfare Concert for the benefit of the Youth Department of the Jewish Community, which took place in Berlin's "Neue Synagoge" located at 30 Oranienburger Strasse. The program included arias sung by the famous tenor Hermann Jadlowker and the Adagio in B-minor for two violins by Johann Sebastian Bach, played by Einstein and the violist Alfred Lewandowski.[20] Early in March 1933, at the end of his second visit to the United States, Einstein became the godfather of Albert, the eight-day-old son of Jacob Landau, the director of the Jewish Telegraphic Agency, at a ceremony in a New York synagogue. During the last two decades of his life, Einstein participated once every two or three years at the discussions that concluded the Friday evening service for Jewish students at Princeton University.

Einstein's last wish was not to be buried in the Jewish tradition, but to be cremated and his ashes scattered, indicating that he disregarded religious rituals until his death on 18 April 1955.

In the late 1940s, reminiscing about his juvenile religious

available. Cf., e.g., R. Highfield and P. Carter, *Private Lives of Albert Einstein*.

[19] P. Frank, *Einstein—His Life and Times*, p. 280.

[20] A photo that shows Einstein on this occasion, playing the violin and wearing a skullcap—as Jews usually do in a synagogue—can be found in W. Cahn, *Einstein—A Pictorial Biography* (Citadel Press, New York, 1955), p. 62.

fervor, Einstein offered a philosophical explanation of his estrangement from traditional religion.

> It is quite clear to me that the religious paradise of youth, which was thus lost, was a first attempt to free myself from the chains of the "merely personal," from an existence which is dominated by wishes, hopes, and primitive feelings. Out yonder there was this huge world, which exists independently of us human beings and which stands before us like a great, eternal riddle, at least partially accessible to our inspection and thinking. The contemplation of this world beckoned like a liberation, and I soon noticed that many a man whom I had learned to esteem and to admire had found inner freedom and security in devoted occupation with it. The mental grasp of this extrapersonal world within the frame of the given possibilities swam as [the] highest aim half consciously and half unconsciously before my mind's eye. Similar motivated men of the present and of the past, as well as the insights which they had achieved, were the friends which could not be lost. The road to this paradise was not as comfortable and alluring as the road to the religious paradise; but it has proved itself as trustworthy, and I have never regretted having chosen it.[21]

Interestingly, Einstein's account does not mention the role that Max Talmey had played in this context. Nor does it describe the emotional struggle and the conscientious conflict that the young Einstein must have experienced

[21]A. Einstein, "Autobiographical Notes," in *Albert Einstein: Philosopher-Scientist*, p. 5.

when he began to doubt the veracity of the Bible. Some biographers see in his religious skepticism the source of his freedom of thought and intellectual independence in scientific reasoning and even regard it as a necessary condition for his discovery of the theory of relativity. Thus, for example, Banesh Hoffmann, who in the thirties had worked on this theory with Einstein for some time and who called Einstein a "creator and rebel," regarded Einstein's "antireligious" stance as the cause of his suspicion of authority. After quoting Einstein's own statement that "to punish me for my contempt for authority, Fate made me an authority myself," Hoffmann declared, "His early suspicion of authority, which never wholly left him, was to prove of decisive importance. For without it, he would not have been able to develop the powerful independence of mind that gave him the courage to challenge established scientific beliefs and thereby revolutionize physics."[22]

Einstein's defiance of authority explains his well-known aversion to social conventions, his nonconformity in apparel and attire, his bohemian style of life during his student years in Zurich, and his friendship and solidarity with colleagues like the Austrian socialist Friedrich Adler or the members of the "Olympia Academy" in Berne, Maurice Solovine, Conrad Habicht, and Michele Angelo Besso. For all of them, the ideologies of Marx and Mach replaced the religion of the Bible. Some authors assign these ideological influences a crucial role in Einstein's intellectual development and regard them, in particular, as the driving force for his creation of the theory of relativity. For example, the sociologist Lewis Samuel Feuer, who in his

[22] B. Hoffmann, with the collaboration of Helen Dukas, *Albert Einstein—Creator and Rebel* (Viking Press, New York, 1972), p. 24.

youth experienced a similar estrangement from religion—though in the different milieu of New York City's Lower East Side—offered his view of how sociopolitical ideas inspired Einstein's theorizing in physics.[23]

> Imagine the youthful genius Einstein in the Zurich setting of a radical student group in which the revolutionary ideas of Marx commingle with those of Mach. Einstein imbibes a notion of the relativity of social laws to transient social systems; the laws of contemporary society are in reality the expressions of bourgeois relations, and are not immutable absolutes. In the evenings, he and Fritz Adler, we may surmise, would argue whether bourgeois observers and socialist ones could describe a common social world, or whether the described social events varied with the social standpoint of the observer; for this too was an issue that Austro-Marxist philosophers debated endlessly. How would this Marxist revolutionary emotion and vision be expressed, though in a sublimated, transfigured form, in the mind of a young "revolutionary genius" in physics? The emotions that gave rise to sociological relativity might then seek to express themselves in a physical relativity; transposed and projected upon the study of the physical world, they would issue in an overthrow of absolute space and time, and in a conception of the relativity of length and time measurements to the observer's state of motion.[24]

[23] L. S. Feuer, "A Narrative of Personal Events and Ideas," in S. Hook, W. L. O'Neill, and R. O'Toole, eds., *Philosophy, History and Social Action* (Kluwer, Dordrecht, 1988), pp. 1–85.

[24] L. S. Feuer, "The Social Roots of Einstein's Theory of Relativity," *Annals of Science* 27 (1971): 277–298, 313–344.

In short, "the emotional standpoint of the Zurich-Berne revolutionary students circle provided the supporting social environment, motivation, and modes of thought for the conception of the theory of relativity."[25] Had Einstein's theory of relativity really been the outcome of such discussions with revolutionary friends, had his congeniality with them been the result of his bohemian style of life and his rejection of authority, and the latter, in turn, the result of his abandonment of his "religious paradise," then, bringing this chain of consequences to its logical conclusion, one would be tempted to contend that what has been called "one of the greatest, perhaps the greatest, of achievements in the history of human thought" had its ultimate *fons et origo* in an estrangement from religion.[26]

Such a conclusion seems to support the thesis that science and religion are irreconcilable antagonists. But Einstein never conceived of the relation between science and religion as an antithesis. On the contrary, he regarded science and religion as complementary to each other or rather as mutually depending on each other, a relation that he described by the metaphor quoted above, "Science without religion is lame, religion without science is blind."[27]

[25] L. S. Feuer, *Einstein and the Generations of Science* (Basic Books, New York, 1974), p. 66.

[26] This laudatory statement was made by Sir Joseph John Thomson, president of the Royal Society of London, on 6 November 1919, when the Fellows of the Royal Society and of the Royal Astronomical Society convened in Burlington House and heard that the British solar eclipse expedition confirmed Einstein's prediction of the deflection of light in a gravitational field. See *New York Times*, 9 November 1919.

[27] A. Einstein, "Science and Religion," Address at the Conference on Science, Philosophy, and Religion, New York, 1940; reprinted in A. Einstein, *Ideas and Opinions* (Crown, New York, 1954, 1982), pp. 44–49. Quotation on p. 46.

In 1930, when interviewed by the Irish writer James Murphy and the Irish mathematician John William Navin Sullivan, Einstein emphatically declared, "I am of the opinion that all the finer speculations in the realm of science spring from a deep religious feeling, and that without such feeling they would not be fruitful."[28]

If, in Einstein's opinion, science and religion are complementary, two questions arise. First, how could he disapprove of religious instruction for his sons on the grounds that it is "contrary to all scientific thinking"?[29] If we exclude the possibility that he changed his mind, then the apparent contradiction can be resolved by recognizing that he used the term "religion" or "religious" in two different senses. In the expression "religious instruction," he used it in the sense of instruction in accordance with a denominational tradition characterized by the rituals of a specific community, whereas in the expression "science without religion," "religion" referred to a pious sentiment of an inspired devotion without any dogmatic indoctrination.[30]

The second question asks whether the thesis—proposed by Feuer, Hoffmann, and others—that Einstein's discovery of the theory of relativity presupposed disrespect of authority and ultimately of religion can be maintained in view of its evident contradiction with Einstein's concept of the relation between science and religion. Some proponents of this thesis regard the very name "theory of relativity" as an argument in support of their point of view,

[28] A. Einstein, "Science and God," *Forum and Century* 83 (1930): 373–379.

[29] P. Frank, *Einstein—His Life and Times*, p. 280.

[30] More precise explications of the term "religion" in its various connotations will engage our attention in chapter 2, which deals with Einstein's writings on the philosophy of religion.

because, they claim, the term "relativity" connotes some latitude or freedom as opposed to absoluteness and therefore invalidates the "absolute sacrosanctity of a religious dogma." "The word 'relativity,' and the expression 'the principle of relativity,'" wrote Feuer, "became emotional symbols of the new generational mode of thought, symbols for the iso-emotional line of generational rebellion."[31] Some may object that the word "relativity" in such interpretations is confused with the term "relativism," which indeed is widely used to denote the denial of the objectivity or absoluteness of ethical or religious values. The philosopher and historian of culture Ernst Cassirer admonished us not to regard the theory of relativity as "a confirmation of the Protagorean doctrine that man is the 'measure of all things.'" He added, "The physical theory of relativity teaches not that what appears to each person is true to him, but, on the contrary, it warns against appearances, which hold only from a particular system."[32]

Moreover, mathematician Felix Klein and physicist Arnold Sommerfeld suggested that the name "theory of relativity" should be replaced by "theory of invariants" because the theory is merely a theory of the invariants of the Lorentz transformation or, in the case of general relativity, of a more general transformation. "The term 'theory of relativity' is an unfortunate choice," wrote Sommerfeld, "its essence is not the relativity of space and time but rather

[31] L. S. Feuer, "The Social Roots of Einstein's Theory," p. 320. "One idea is iso-emotional with another, or with any cultural manifestation, when it is an expression, reflection, outcome, or projection of the same sort of emotion." Ibid., p. 315.

[32] E. Cassirer, *Substance and Function—Einstein's Theory of Relativity* (Open Court, London, 1963), p. 392.

the independence of the laws of nature from the viewpoint of the observer. The bad name has misled the public to believe that the theory involves a relativity of ethical conceptions, somehow like Nietzsche's *Beyond Good and Evil*."[33]

In fact Einstein never made these associations with relativity, nor was he the first to use the term relativity in physics. The adjective "relative" (Latin: *relativus*) has, of course, a history reaching back to antiquity. Newton applied it in the first Scholium of his *Principia* when he distinguished between *tempus absolutum* and *tempus relativum, spatium absolutum* and *spatium relativum*, and *motus absolutus* and *motus relativus*. But the noun "relativity" or its equivalent in other languages was first used in the nineteenth century by poets, such as Samuel Taylor Coleridge, and by philosophers, such as John Stuart Mill, mostly in the expression "relativity of knowledge." Einstein probably encountered this term for the first time when as a student he read the first volume of Jules Violle's *Lehrbuch der Physik* (1892) and Henri Poincaré's *La Science et Hypothèse* (1902), in which the term "le principe de la relativité" denotes the statement that "the motion of an arbitrary system must obey the same laws, whether referred to fixed axes or to moving axes undergoing a uniform rectilinear motion." Clearly, Poincaré, in every respect a conservative, can hardly be assumed to have linked any ideological connotation with this term.[34]

Einstein himself once emphasized, "In the relativity the-

[33] A. Sommerfeld, "Philosophie und Physik seit 1900," *Naturwissenschaftliche Rundschau* 1 (1948): 97–100; *Gesammelte Schriften* (Vieweg, Braunschweig, 1968), vol. 4, pp. 640–643.

[34] For details on the use of this term, see *The Collected Papers of Albert Einstein* (Princeton University Press, Princeton, N.J., 1989), vol. 2, p. 259.

ory, it is no question of a revolutionary act but of a natural development of lines which have been followed for centuries."[35] On another occasion he declared that this theory "grew out of the Maxwell-Lorentzian electromagnetics as a surprisingly simple summary and generalization of previously independent hypotheses."[36] Obviously, a "simple summary and generalization" of previous ideas has nothing to do with a revolt against authority or religion.

Einstein himself repeatedly insisted that his theory of relativity should not be regarded as a revolutionary break with the past. Thus, on April 2, 1921, when he arrived in New York on his first visit to the United States and was interviewed by reporters of the *New York Times*, Einstein declared, "There has been a false opinion widely spread among the general public that the theory of relativity is to be taken as differing radically from the previous developments in physics. . . . The men who have laid the foundations of physics on which I have been able to construct my theory are Galileo, Newton, Maxwell, and Lorentz."[37] He often called his theory "simply a systematic development of the electrodynamics of Maxwell and Lorentz," and regarded it as an "evolution," not a revolution, of the science of dynamics.[38]

[35] C. Seelig, *Albert Einstein—A Documentary Biography* (Staples Press, London, 1956), p. 82.

[36] A. Einstein, *Über die spezielle und die allgemeine Relativitätstheorie* (Vieweg, Braunschweig, 1920), p. 28.

[37] *New York Times*, 4 April 1921. Also quoted in G. Holton, "Einstein's search for the Weltbild," *Proceedings of the American Philosophical Society* 125 (1981): 1–15.

[38] A. Einstein, "What is the theory of relativity?" *London Times*, 28 November 1919; reprinted in A. Einstein, *Ideas and Opinions* (Crown Publishers, New York, 1954, 1982), pp. 227–232; p. 248.

A valuable but little known source of information on Einstein's way of thinking is the diary of Count Harry Kessler, a prominent diplomat, art connoisseur, litterateur, and socialite of Berlin's intellectual circles in the twenties. Because he wrote everything down immediately after it had happened, the records in his diary are highly reliable. Kessler met frequently with Einstein at receptions, dinners, and other occasions. At a reception given by the Einsteins on March 20, 1922, the discussion veered to the theory of relativity and the question of how far it differs from classical theories in physics. "It is the inextricable connection between matter, space, and time that is new in the theory," said Einstein. And Kessler's entry in his diary continues:

> What he does not understand is why people have become so excited about it. When Copernicus dethroned the earth from its position as the focal point of creation, the excitement was understandable because a revolution in all man's ideas did occur. But what change does his own theory produce in humanity's view of things? It is a theory which harmonizes with every reasonable outlook or philosophy and does not interfere with anybody being an idealist or materialist, pragmatist, or whatever else he likes.[39]

In his profound study of the conceptual development of the theory of relativity, Gerald Holton not only confirms Einstein's judgment of the nonrevolutionary character of the theory, he even regards it as an example of the general rule that a "so-called scientific 'revolution' turns out to be

[39] Harry Graf Kessler, *Tagebücher 1918–1937* (Insel Verlag, Frankfurt, 1961); *The Diary of a Cosmopolitan* (Weidenfeld and Nicolson, London, 1971), p. 157.

at bottom an effort to return to a classical purity."[40] This is, of course, the very antithesis of the thesis of Feuer and Hoffmann, which has also been rejected by John L. Heilbron in his review of Feuer's book.[41] Paul Forman, who is well-known for his studies on the sociopolitical conditions of the development of modern physics, came to the conclusion that "it would be a mistake to claim that Einstein and his close friends belonged to an alternative culture. They were not science-oriented bohemians, nor were they social revolutionaries."[42]

In any case, there can be little doubt that the predominant motivations that led Einstein to his development of the theory of relativity were purely physical considerations, like his recognition that Maxwell's electrodynamics, as understood at the beginning of the century, "when applied to moving bodies, leads to asymmetries which do not appear to be inherent in the phenomena,"[43] or his sudden realization that the important concept of simultaneity must be defined in terms of physical operations.[44] Einstein was

[40] G. Holton, "On the origins of the special theory of relativity," *American Journal of Physics* 28 (1960): 627–636; reprinted in G. Holton, *Thematic Origins of Scientific Thought: Kepler to Einstein* (Harvard University Press, Cambridge, Mass., 1973), pp. 165–183.

[41] J. L. Heilbron, book review, *Science* 185 (1974): 777–779.

[42] P. Forman, Introduction to L. Infeld, *Why I Left Canada: Reflections on Science and Politics* (McGill-Queen's University Press, Montreal, 1978), p. 9.

[43] A. Einstein, "Zur Elektrodynamik bewegter Körper," *Annalen der Physik* 17 (1905): 891–921; H. A. Lorentz, A. Einstein, H. Minkowski, and H. Weyl, *The Principle of Relativity* (Dover, New York, 1952), pp. 35–65; *The Collected Papers of Albert Einstein*, vol. 2 (Princeton University Press, Princeton, N.J., 1989), pp. 276–306.

[44] "An analysis of the concept of time was my solution." A. Einstein, "How I created the theory of relativity," lecture given in Kyoto, 14 December 1922. *Physics Today* 35 (August 1982): 45–47.

fully aware, however, that the construction of a new far-reaching theory, even if motivated by purely physical considerations, is not an instantaneous mental process. Thus, he began his Kyoto lecture, in which he described the impact of his analysis of time, with the declaration: "It is not easy to talk about how I reached the idea of the theory of relativity; there were so many hidden complexities to motivate my thought, and the impact of each thought was different at different stages in the development of the idea. I will not mention them all here." In a later statement, he recalled the above-mentioned symmetry problem and said, "What led me more or less directly to the special theory of relativity was the conviction that the electromagnetic force acting on a body in motion in a magnetic field was nothing but an electric field." He immediately added, "There is, of course, no logical way leading to the establishment of a theory but only groping constructive attempts by careful considerations of factual knowledge."[45]

In referring to "hidden complexities" or "groping constructive attempts," was Einstein alluding to the possibility that extraphysical considerations had been involved? As shown below, if there had been such extraphysical ingredients in Einstein's construction of his theory of relativity they could not have been sociological or political, as has been contended by Feuer, for example, but they could only have been philosophical or, perhaps, religious, in the sense of Einstein's definition of this term.

[45] Statement sent to a special meeting of the Cleveland Physical Society, 19 December 1952, honoring the centenary of Michelson's birth; printed in R. S. Shankland, "Michelson-Morley Experiment," *American Journal of Physics* 32 (1964): 16–35.

The exact formulation of Einstein's definition of religion is, of course, part of his philosophy of religion and will be considered in chapter 2. At present it suffices to know what he meant by "being religious." In his diary, Count Kessler tells of a dinner that took place at publisher Samuel Fischer's home in Berlin February 14, 1927. Apart from Einstein and Kessler, the famous novelist Gerhart Hauptmann and the well-known Berlin critic Alfred Kerr were guests. Pretending to be a firm believer in astrology, Hauptmann asked Einstein whether he shared this belief. Einstein, who had just read Lucien Levy-Bruhl's book *Die geistige Welt der Primitiven* about the demonology of early cultural levels and its effects on ancient religious beliefs, told Hauptmann that faith in astrology evolved from an ancient belief in demons. Einstein, who did not believe in such supernatural beings, strongly condemned astrology as a superstition.

The conversation then turned from astrology to religion. "Kerr," Kessler reported,

> constantly interrupted with facetious remarks . . . the subject of God was a special butt for his derision. I tried to silence him and said that, since Einstein is very religious, he should not needlessly hurt his feelings. "What?" exclaimed Kerr, "It isn't possible! I must ask him right away. Professor! I hear that you are supposed to be deeply religious?" Calmly and with great dignity, Einstein replied, "Yes, you can call it that. Try and penetrate with our limited means the secrets of nature and you will find that, behind all the discernible concatenations, there remains something subtle, intangible and inexplicable. Veneration for this force

beyond anything that we can comprehend is my religion. To that extent I am, in point of fact, religious."[46]

LET US now discuss briefly the philosophical background of Einstein's scientific work or, more precisely, how far philosophy has influenced his physics and, inversely, how far his achievements in physics have affected his philosophical outlook. Both issues are closely interrelated.[47] Einstein himself was aware of this interdependence when he wrote to Cornelius Lanczos, "I began with a skeptical empiricism more or less like that of Mach. But the problem of gravitation converted me into a believing rationalist, that is, into someone who searches for the only reliable source of Truth in mathematical simplicity."[48]

By "the problem of gravitation," he meant the general theory of gravitation. He explained his initial endorsement of "skeptical empiricism," according to which no knowledge with existential reference is possible independent of experience, as the result of having read the writings of Hume and of Mach. He acknowledged repeatedly that the empiricism of Hume and Mach had deeply influenced his early work on relativity. Thus, he wrote to Carl Seelig, "The critical thought necessary for the discovery of this central point [i.e., the recognition of the need for an operational definition of the concept of distant simultaneity] was

[46] H. G. Kessler, *The Diary of a Cosmopolitan*, p. 322.

[47] The second issue will be discussed in general—i.e., without being restricted to Einstein personally—in chapter 3, because it touches on the implications of Einstein's theories for religious thought.

[48] Einstein to C. Lanczos, 24 January 1938. Einstein Archive, reel 15-267.

afforded me decisively by the reading of David Hume's and Ernst Mach's philosophical writings." On another occasion, he declared that he had studied Hume's *Treatise of Human Nature* "with fervor and admiration shortly before the discovery of the theory of relativity." "It is very well possible," he added, "that without these philosophical studies I would not have arrived at the special theory of relativity."[49]

Incidentally, Einstein's statement sharply contradicts David Hilbert's explanation of how Einstein discovered relativity. Hilbert, the eminent Goettingen mathematician who preceded Einstein by five days in presenting the famous field equations of general relativity, once asked a gathering of mathematicians, "Do you know why Einstein said the most original and profound things about space and time that have been said in our generation? Because he had learned nothing about all the philosophy and mathematics of time and space."[50]

Hilbert's explanation is also contradicted by Max Talmey's statement that a teenage Einstein had studied Immanuel Kant's *Critique of Pure Reason*.[51] As is well known, Kant claimed to have proved that space and time do not subsist as entities in themselves but are rather a priori forms of intuition and, as such, preconditions for the possibility of experience. Einstein's work in physics convinced him that Kant's differentiation between a priori and a pos-

[49] C. Seelig, *Albert Einstein—A Documentary Bibliography*, p. 67.

[50] P. Frank, *Einstein—His Life and Times*, p. 206.

[51] M. Talmey, *The Relativity Theory Simplified*. See also C. Seelig, *Albert Einstein*, p. 14, where it is stated that "the 16-year-old youth intoxicated himself with Kant's *Critique of Pure Reason*."

teriori or empirical notions "is erroneous, i.e., does not do justice to the problem in a natural way. All concepts, even those which are closest to experience, are from the point of view of logic freely chosen conventions, just as is the case with the concept of causality, with which this problematic concerned itself in the first instance."[52]

In a letter to Max Born in 1918, Einstein wrote: "Once you concede to him [Kant] merely the existence of synthetic a priori judgements, you are trapped. I have to water down the 'a priori' to 'conventional,' so as not to have to contradict him, but even then the details do not fit. Anyway it is very nice to read, even if it is not as good as his predecessor Hume's work. Hume also had a far sounder instinct."[53] Still, there were issues on which Einstein agreed with Kant. Referring to the fact that the totality of our sense experiences can be put in order by means of thinking, a fact "which leaves us in awe, but which we shall never understand," Einstein said that "the eternal mystery of the world is its comprehensibility." He declared, "It is one of the great realizations of Immanuel Kant that the setting up of a real external world would be senseless without this comprehensibility."[54]

Talmey's comment that "Kant became Albert's favorite philosopher after he had read through his *Critique of Pure Reason* and the work of other philosophers," if correct at all, could have referred only to the young Einstein. By 1920 at

[52] P. A. Schilpp, ed., *Albert Einstein: Philosopher-Scientist*, p. 13.

[53] *The Born-Einstein Letters* (Macmillan, London, 1971), p. 7.

[54] A. Einstein, "Physics and Reality," *Journal of the Franklin Institute* 221 (1936): 349–382; reprinted in *Out of My Later Years* (Littlefield, Adams & Co., Totowa, N.J., 1967), pp. 58–94.

least, the philosopher whom Einstein admired most was Baruch (later, Benedictus) Spinoza, the seventeenth-century Jewish philosopher, who was excommunicated by the Amsterdam synagogue and declined the Heidelberg professorship in order to live as a lens grinder, leading an independent life dedicated to philosophical reflections. Einstein already had studied Spinoza's *Ethics* in Berne with his friends of the Olympia Academy and resumed this study several years later.[55] His earliest recorded references to Spinoza date from 1920. In that year he composed a poem entitled "Zu Spinozas Ethik" (see the appendix for the entire poem in the original German). It begins with the following words,

> How much do I love that noble man
> More than I could tell with words
> I fear though he'll remain alone
> With a holy halo of his own.

Some background may be helpful so that the reader not conversant with Spinoza's philosophy can understand to what extent Einstein concurred with him. Rejecting the traditional theistic concept of God, Spinoza denied the existence of a cosmic purpose on the grounds that all events in nature occur according to immutable laws of cause and effect. The universe is governed by a mechanical or mathematical order and not according to purposeful or moral intentions. Though he employed the notion of "God," Spinoza applied it only to the structure of the impersonal cosmic order and declared

[55] Einstein apparently used the German translation of Spinoza's *Ethica ordine geometrico demonstrata*, published by F. Meiner, Leipzig, in 1910, his copy of which is now part of the Einstein estate at the Hebrew University in Jerusalem.

that "neither intellect nor will appertain to God's nature." He therefore denied the Judeo-Christian conception of a personal God. What the Bible refers to as divine activities are identified by Spinoza with the lawlike course of nature. God is the "infinite substance" having the attributes of extension and thought. God is devoid of ethical properties, for good and evil are only relative to human desires. What is commonly called "God's will" is identical with the laws of nature. People do not act freely in the sense of having alternatives to their actions; their belief in freedom arises only from their ignorance of the causes of the desires that motivate their actions. The ultimate object of religious devotion can only be the perfect harmony of the universe, and human aspirations must accept the inexorable dictates of the deterministic laws that govern life.

Evidently, Einstein was enchanted by Spinoza's *Ethics*, but he never considered himself an expert on Spinoza's writings. In 1932, the tercentenary of Spinoza's birth, Einstein was asked by several people to write about Spinoza but refused. For example, Siegfried Hessing, a publicist from Czernowits, Rumania, invited him to join Henri Bergson, Sigmund Freud, Stefan Zweig, Romain Rolland, and others in writing a series of essays in honor of Spinoza. Einstein replied, "Unfortunately, to love Spinoza does not suffice to be allowed to write about him; this one must leave to those who have gone further into the historical background."[56] When asked by Dr. Dagobert Runes, a New York book publisher, to write a short essay on "the ethical significance of Spinoza's philosophy," he declined the invi-

[56] Einstein to S. Hessing, 8 September 1932. Einstein Archive, reel 33-288.

tation on the following grounds, "I do not have the professional knowledge to write a scholarly article about Spinoza. But what I think about this man I can express in a few words. Spinoza was the first to apply with strict consistency the idea of an all-pervasive determinism to human thought, feeling, and action. In my opinion, his point of view has not gained general acceptance by all those striving for clarity and logical rigor only because it requires not only consistency of thought but also unusual integrity, magnanimity and—modesty."[57]

Obviously, it was not so much Einstein the physicist as Einstein the philosopher who admired Spinoza. Any attempt to explain his veneration of Spinoza by claiming that the *Ethics* somehow anticipated Einstein's scientific thought, that Spinoza's notion of "substance" ("substantia") or its attribute "extension" ("extension" as used, e.g., in proposition 2 of part 2 of the *Ethics*) anticipated the concept of space-time as used in the special or general theory of relativity, is artificial and unwarranted for it ignores the historical context of these notions.[58] The only connecting link between Spinoza's philosophy and Einstein's physics and philosophy is the idea of an unexceptionable determinism, which, as seen below,

[57] "Spinoza ist der Erste gewesen, der den Gedanken der deterministischen Gebundenheit alles Geschehens wirklich konsequent auf das menschliche Denken, Fühlen und Handeln angewendet hat. Nach meiner Ansicht hat sich sein Standpunkt unter den um Klarheit und Folgerichtigkeit kämpfenden nur darum nicht allgemein durchsetzen können, weil hierzu nicht nur konsequenz des Denkens, sondern auch eine ungewöhnliche Lauterkeit, Seelengrösse und—Bescheidenheit gehört." Einstein to D. Runes, 8 September 1932. Einstein Archive, reel 33-286.

[58] For a thorough critique of such misinterpretations, see M. Paty, "Einstein and Spinoza," in M. Grene and D. Nails, eds., *Spinoza and the Sciences* (Reidel, Dordrecht, 1986), pp. 267–302.

decisively influenced Einstein's religious credo. Einstein also greatly admired Spinoza's lack of ego, his flight from the "merely personal" throughout his writings. The separation of the excommunicated Jew from his family and home also contributed to Einstein's sympathy for Spinoza. In spite of his unprecedented fame and international adulation, Einstein ultimately remained, as he called himself, a "lone traveler: I have never belonged to my country, my home, my friends, or even my immediate family, with my whole heart. . . . I have never lost a sense of distance and a need for solitude," he confessed in 1930.[59]

Einstein felt akin to Spinoza because he realized that they shared a need for solitude as well as the fate of having been reared within the Jewish heritage but having become subsequently alienated from its religious tradition. Einstein's opinion about the relation between Judaism and Spinozism can be gathered from his correspondence with Willy Aron, the author of a book on Spinoza. "Although I firmly believe," wrote Einstein, "that the chasm between Jewish theology and Spinozism can never be bridged, I am not less convinced that Spinoza's contemplation of the world ("Weltanschauung") was thoroughly imbued with the principles and sentiments that characterize so many Jewish intellectuals. I feel I would never have come so near to Spinoza had I not myself been of Jewish extraction and grown up in a Jewish milieu."[60] In a similar vein, the late Harry Austryn Wolfson, professor of Jewish philosophy at Harvard University wrote in his important treatise on

[59] A. Einstein, "What I Believe," *Forum and Century 84*, 193–194 (1930); reprinted in *Ideas and Opinions*, pp. 8–11.

[60] Einstein to W. Aron, 14 January 1943. Einstein Archive, reel 33-296.

Spinoza's philosophy, "We cannot get the full meaning of what Benedictus says unless we know what has passed through the mind of Baruch."[61]

Einstein was most influenced by Spinoza's thesis of an unrestricted determinism and the belief in the existence of a superior intelligence that reveals itself in the harmony and beauty of nature. In any case, these were the interpretations that Einstein gave to Proposition 29 in the first part of Spinoza's *Ethics*: "In rerum natura nullum datur contingens, sed omnia ex necessitate divinae naturae determinata sunt ad certo modo existendum, et corporandum" [In the nature of things nothing is contingent but all things are determined by the necessity of divine nature existing and operating in a certain mode], and to the expression "divina natura" or "deus sive natura," respectively. Unrestricted determinism, Einstein argued, does not admit a "God who rewards and punishes the objects of his creation and whose purposes are modeled after our own."

Like Spinoza, Einstein denied the existence of a personal God, modeled after the ideal of a superman as we would say today. In accordance with Jewish thought, both Einstein and Spinoza conceived of God as an abstract entity in accordance with the biblical "Thou shalt not make unto thee a graven image, or any likeness of any thing" (Exodus 20:4) and in accordance with Maimonides' *Third Principle of Faith*, "I firmly believe that . . . no bodily accidents apply to Him, and that there exists nothing whatever [that] resembles Him."[62]

[61] H. A. Wolfson, *The Philosophy of Spinoza* (Harvard University Press, Cambridge, Mass., 1934), vol. 1, p. vii.

[62] See, e.g., P. Birnbaum, ed., *Daily Prayer Book* (Hebrew Publishing, New York, 1949), p. 154.

When Einstein was once asked to define God, he gave the following allegorical answer,

> I'm not an atheist, and I don't think I can call myself a pantheist. We are in the position of a little child entering a huge library filled with books in many languages. The child knows someone must have written those books. It does not know how. It does not understand the languages in which they are written. The child dimly suspects a mysterious order in the arrangement of the books but doesn't know what it is. That, it seems to me, is the attitude of even the most intelligent human being toward God. We see the universe marvelously arranged and obeying certain laws but only dimly understand these laws. Our limited minds grasp the mysterious force that moves the constellations. I am fascinated by Spinoza's pantheism, but admire even more his contribution to modern thought because he is the first philosopher to deal with the soul and body as one, and not two separate things.[63]

At about the same time, in April 1929, Cardinal O'Connell, Archbishop of Boston, admonished the members of the New England Catholic Club of America not to read anything about the theory of relativity, because it is a "befogged speculation producing universal doubt about God and his Creation . . . cloaking the ghastly apparition of atheism."[64]

Worried by the Archbishop's exprobration, Rabbi Herbert

[63] G. S. Viereck, *Glimpses of the Great* (Macauley, New York, 1930), quoted by D. Brian, *Einstein—A Life*, p. 186.

[64] *New York Times*, 25 April 1929, p. 60.

S. Goldstein of the Institutional Synagogue in New York cabled Einstein, "Do you believe in God? Stop. Prepaid reply fifty words." Einstein replied, "I believe in Spinoza's God who reveals himself in the orderly harmony of what exists, not in a God who concerns himself with fates and actions of human beings." Rabbi Goldstein commented that this reply

> very clearly disproves . . . the charge of atheism made against Einstein. In fact, quite the reverse is true. Spinoza, who is called "the God-intoxicated man" and who saw God manifest in all of nature, certainly could not be called an atheist. . . . Einstein's theory, if carried out to its logical conclusions would bring mankind a scientific formula for monotheism. He does away with all thought of dualism or pluralism. There can be no room for any aspect of polytheism.[65]

Chapman Cohen, president of the National Secular Society in England, an association mostly of freethinkers, devoted a whole chapter of his book, *God and the Universe*, to his claim that this communication between Goldstein and Einstein actually led to an affirmation of atheistic ideology. "The portraits we have seen of Einstein," Cohen wrote, "show him to be not destitute of humour, and we fancy he must have felt he was doing a little 'leg-pulling' when he gave his answer to Rabbi Goldstein."

Einstein's declaration that he believes in the God of Spinoza can be of no use to anybody who is religious. If God, according to Einstein, is not concerned with the actions and prayers of man, Cohen continued, it is obviously of no use to pray to him. "One might as well pray to the Albert Memo-

[65] Ibid.

rial. . . . What significance have all the churches, synagogues, mosques, and other gathering places of the religiously afflicted if they are worshipping a God who takes no interest in their fates or their actions." Einstein's confession is but a confession of "practical atheism," because there is no difference between there being no God to bother about man, and there being a God who does not concern himself with the fates and actions of human beings.

Spinoza's God is thoroughly deterministic, and, "if one translates his ideas into modern terms, completely atheistic." Goldstein's praise of Einstein's reply as "a scientific formula for monotheism" only shows that "we have reached the stage where genuine religion finds it increasingly hard to live honestly, and altogether lacks the courage to die with courage and dignity. Anything will do, so long as it is given the name of God. It is still a term which exerts a hypnotic power over the unthinking, and it is by the support of the unthinking that established religion today hopes to carry on." Cohen concluded this chapter with the remark that "one can imagine the twinkle in the eyes of Albert Einstein when he replied to the Rabbi's inquiry, 'I believe in Spinoza's God.' Perhaps he whispered to himself, 'And that is no god at all.'"[66]

But Einstein always made a sharp distinction between his disbelief in a personal God and atheism. Not long after he had cabled his answer to Rabbi Goldstein, he received from Eduard Büsching of Stuttgart a copy of Büsching's book, entitled *Es gibt keinen Gott* [*There Is no God*], published under the pseudonym Karl Eddi.[67] This book defines

[66] C. Cohen, *God and the Universe*, 3d ed. (Pioneer Press, London), pp. 126–132.

[67] K. Eddi, *Es gibt keinen Gott—Bekenntnisse eines Unbekannten* (Koch, Neff & Oetinger, Stuttgart, 1929).

religion as "the abortive attempt, roused by deference and fear of the unknown, to establish a direct and personal relation to an imaginary superior being God or Gods, resembling mankind and ruling it, but not existing in reality." It concludes with the statement, "Where science grows, religion wanes; where religion thrives, science withers,"[68] the exact antithesis of Einstein's statement "Science without religion is lame, religion without science is blind."[69]

Einstein courteously responded that the book did not deal with the notion of God but only with that of a personal God and therefore should be called *Es gibt keinen persönlichen Gott*. He continued,

> We followers of Spinoza see our God in the wonderful order and lawfulness of all that exists and in its soul ("Beseeltheit") as it reveals itself in man and animal. It is a different question whether belief in a personal God should be contested. Freud endorsed this view in his latest publication. I myself would never engage in such a task. For such a belief seems to me preferable to the lack of any transcendental outlook of life, and I wonder whether one can ever successfully render to the majority of mankind a more sublime means in order to satisfy its metaphysical needs.[70]

Einstein, as we see, was far from disputing the usefulness of religious education; he objected to it, as he had for his children, only when he suspected that the major objec-

[68] "Wo Wissen Macht, da stirbt der Glaube, wo der Glaube herrscht, verweht der Geist." Ibid., p. 73.

[69] A. Einstein, "Science and Religion."

[70] Einstein to E. Büsching, 25 October 1929. Einstein Archive, reel 33-275.

tive was to teach religious ceremonies or formal rituals instead of the development of ethical values. Einstein's conception of the relation between religion and ethics is closely analyzed in chapter 2. According to Einstein, even science at an advanced stage, cannot define, let alone commend, ethical values. For science is confined to what *is* and ethics to what *should be*, and no path leads from the knowledge of what is to the knowledge of what should be.

In 1930, Einstein was invited by the *New York Times* to contribute an essay on his conception of the relation between science and religion. In this article, entitled "Religion and Science,"[71] Einstein used, apparently for the first time, the term "cosmic religious feeling" to describe the emotional state that one experiences when one recognizes the "futility of human desires and the sublimity and marvelous order which reveals itself both in nature and in the world of thought." In assuming *one* order in nature and in thought, Einstein followed, consciously or not, Spinoza's doctrine: "Ordo et connexio idearum idem est, ac ordo et connexio rerum."[72] This Spinozistic tenet underlies Einstein's epistemological realism, his belief that a rational explanation of the universe is possible, his belief in the "mysterious comprehensibility of the world." It explains, for example, Einstein's reaction to Eddington's cable containing the re-

[71] A. Einstein, "Religion and Science," *New York Times Magazine*, 9 November 1930, section 5, pp. 1–4; German translation in *Berliner Tageblatt*, 11 November 1930, pp. 1–3; reprinted as the title essay in *Cosmic Religion with Other Opinions and Aphorisms* (Covici-Friede, New York, 1931), pp. 43–54; also in *The World as I See It* (Philosophical Library, New York, 1949), pp. 24–28; and in *Ideas and Opinions* (Crown, New York, 1954, 1982), pp. 36–40.

[72] "Order and connection of ideas is the same as order and connection of things." B. Spinoza, *Ethica*, Proposition 7, part 2.

sults of the 1919 expedition to measure the deflection of light in a gravitational field. On receiving the cable, Einstein's assistant Ilse Rosenthal-Schneider expressed her joy that these astronomical observations confirm the general theory of relativity, and Einstein reportedly said, "But I knew that the theory was correct." When she asked him, "What if there had been no confirmation of the predictions?" he countered, "Then I would have been sorry for the dear Lord—the theory is correct."[73]

However, when Einstein later applied this parallelism between "ordo idearum" and "ordo rerum" in his study of quantum mechanics, his insistence on the *primacy* of an unrestricted determinism somewhat abated. In fact, as Wolfgang Pauli wrote in 1954 to Max Born, "Einstein does not consider the concept of 'determinism' to be as fundamental as it is frequently held to be (as he told me emphatically many times). . . . [H]e disputes that he uses as a criterion for the admissibility of a theory the question: 'Is it rigorously deterministic?'"[74]

This shift in Einstein's position was, partially at least, the result of his failure to disprove the Heisenberg indeterminacy relations, which form an integral part of the stan-

[73] I. Rosenthal-Schneider, "Reminiscences of Conversation with Einstein," 23 July 1957; "Reminiscences of Einstein," in H. Woolf, ed., *Some Strangeness in the Proportion* (Addison-Wesley, Reading, Mass., 1980), p. 523.

[74] "Insbesondere hält Einstein (wie er mir ausdrücklich wiederholte) den Begriff 'Determinismus' nicht für so fundamental wie es oft geschieht und leugnete energisch . . . das er 'als Kriterium für eine zulässige Theorie' die Frage benutzt: 'ist sie streng deterministisch?'" Letter from W. Pauli to M. Born, dated 31 March 1954. *Albert Einstein—Hedwig und Max Born, Briefwechsel* (Nymphenburger Verlagshandlung, München, 1969), p. 293; *Born-Einstein Letters*, p. 221.

dard version of quantum mechanics. These relations deny the ascertainability of measuring the exact initial values of canonically conjugate observables, such as the position and the momentum of a particle, which are necessary for the prediction of the future state of the system and thus deprive the notion of determinism of any physical meaning. Even if they do not imply the possibility of proving the nonexistence of determinism, they imply at least the impossibility of proving the existence of determinism.

The principle that Einstein, as a consequence of his critique of quantum mechanics, considered even more fundamental than the requirement of determinism, was called by him the "Trennungsprinzip" (principle of separation).[75] It demands that the outcome of a measurement performed on a physical system cannot depend on the outcome of a measurement performed simultaneously on another system that is spatially separated from the first one; or, briefly expressed, it denies the possibility of an immediate interaction between spatially separated systems. How Einstein arrived, via the famous 1935 "Einstein-Podolsky-Rosen incompleteness argument," at this principle and why he attached such importance to it is explained in chapter 3 in the context of certain claims for a theological significance of this principle.

Compared with Einstein's rejection of his earlier endorsement of Mach's positivism in favor of a rational realism as a result of his work on general relativity, the present partial demotion of determinism in favor of the principle of separation was, of course, only a minor change in his philosophy of science.[76] Moreover, because this shift did

[75] The term "Trennungsprinzip" was used in Einstein to E. Schrödinger, 9 June 1935. Einstein Archive, reel 22-047.

[76] For Einstein's gradual disengagement from Mach's positivism,

not imply a denial of determinism, its effect on Einstein's conception of religion cannot be expected to have been very serious. Indeed, in 1935 and thereafter, Einstein did not revise his former religious conviction. It may be significant, however, that, although most of his writings about religion and its relation to science date to the period from 1930 to 1935, his interest in this subject after that period—that is, after the publication of the Einstein-Podolsky-Rosen paper—seems to have waned; only occasionally was it rekindled by epistolatory inquiries.

DID EINSTEIN'S conception of religion or his religious sentiments affect his scientific work? Two questions are involved. First, was his religiosity a psychological or spiritual driving force that stimulated him to endure the hardships of concentrated work, sometimes under quite difficult physical conditions? Second, did his religious conceptions affect the very substance of his work; in other words, was the content of his physical theories influenced by what he called his cosmic religion?

Einstein himself answered the first question—though not with respect to himself but with respect to other great physicists—when he declared, "What a deep conviction of the rationality of the universe [the Spinozistic-Einsteinian expression for religiosity] . . . Kepler and Newton must have had to enable them to spend years of solitary labor in disentangling the principles of celestial mechanics!"[77]

Einstein never said that his religious feelings strengthened his capability to work, unless we interpret his dictum

---

see G. Holton, "Mach, Einstein, and the search for reality," *Daedalus* 97 (1968): 636–673, reprinted in Holton, *Thematic Origins of Scientific Thought*, pp. 219–259.

[77] A. Einstein, *Ideas and Opinions*, p. 39.

"Science without religion is lame" in that way.[78] If we recall that, for him, music was an expression of religious feeling and that often, while playing music, he "suddenly" found the solution to a scientific problem that had intrigued him for some time, then a positive answer to the first question cannot be totally discarded. As far as we know, the last few days before completing the general theory of relativity probably encompassed the most concentrated work of his life. A vivid description of those days, reported by his wife Elsa, can be found in Charles Chaplin's autobiography:

> The Doctor came down in his dressing gown as usual for breakfast but he hardly touched a thing. I thought something was wrong, so I asked what was troubling him. "Darling," he said, "I have a wonderful idea." And after drinking his coffee, he went to the piano and started playing. Now and again he would stop, making a few notes then repeat: "I've got a wonderful idea, a marvelous idea!" I said: "Then for goodness' sake tell me what it is, don't keep me in suspense." He said: "It's difficult, I still have to work it out."
>
> She told me he continued playing the piano and making notes for about half an hour, then went upstairs to his study, telling her that he did not wish to be disturbed, and remained there for two weeks. "Each day I sent him up his meals," she said, "and in the evening he would walk a little for exercise, then return to his work again. Eventually," she said, "he came down from his study looking very pale. 'That's it,' he told me, wearily putting two sheets of paper on the table. And that was his theory of relativity."[79]

[78] A. Einstein, "Science and Religion."

[79] C. Chaplin, *My Autobiography* (Bodley Head, London, 1964), pp.

If his religious sentiments, either directly or through their expression by music, gave him the strength and enthusiasm to work so strenuously in developing his general theory of relativity, then these feelings certainly also motivated his indefatigable tenacity in searching for a unified field theory, a task on which he embarked soon after the completion of his general theory. His aim, as he described it once to his former student Fritz Zwicky, was "to obtain a formula that will account in one breath for Newton's falling apple, the transmission of light and radio waves, the stars, and the composition of matter." As is well known, he did not succeed, but in spite of innumerable disappointments, he never ceased to believe that there *ought* to exist such a theory. This belief may well have been rooted in his Spinozistic conviction in the unity of nature: "Deum unicum, hoc est in rerum natura non nisi unam substantiam dari." [God is One, hence in the nature of things only one substance is given; *Ethics*, corollary 1 to proposition 14, part 1]. Spinoza taught that nature is divine and God is One, and the most fundamental maxim of Judaism, the "Shma' Israel" ("Hear, O Israel, the Lord is our God, the Lord is One"; Deuteronomy 6:4) was well known to Einstein from his early religious instruction.[80] Clearly Einstein's indomitable striving throughout his later lifetime for "oneness" in physics provides a positive answer to both questions

346–347. Elsa gave this report early in 1931 at Chaplin's Beverly Hills home during a dinner party to which she and Albert had been invited by Chaplin after a sightseeing tour of Los Angeles. It was Einstein's second visit to America, spent mainly at the California Institute of Technology in Pasadena.

[80] The oneness of God is also declared in the New Testament, I *Corinthians* 1:8: "There is no God but one," and in the Koran, *Sura* 112: "There *is* God, the One and Only."

posed above; it accounts for his self-devotion to his work as well as for the substance of his work.

We can see that the answer to the second question is positive in other respects, apart from Einstein's search for a unified field theory. For Einstein's religious conviction, following Spinoza, was based on the assumption of an unrestricted determinism, according to which, not only the motions of massive gravitating bodies, such as the stars, but also atomic processes are ruled by strict deterministic laws. Hence, Einstein's persistent objection to the new quantum mechanics, on the grounds that "God does not play at dice," was, at least to some extent, religiously motivated. Others, for example, Cornelius Lanczos, who had been working with Einstein for some time in Berlin, and Georg Herz Shikmoni, the chairman of the Spinozaeum in Haifa, claimed that certain specific physical ideas in the theory of relativity were influenced by religious considerations.[81] Shikmoni even declared that Einstein's famous mass-energy relation, usually expressed by the formula $E = mc^2$, corresponds to a proposition in the *Ethics*.[82]

All those mentioned who believed that Einstein's theory of relativity was to some extent religiously influenced or motivated were convinced that such motives did not impair but rather enhanced the development of the theory without damaging its scientific importance. The opposite

[81] C. Lanczos, *Judaism and Science* (Leeds University Press, Leeds, 1970).

[82] "Ich möchte in diesem Zusammenhang auf die Verwandschaft der Einsteinformel $E = m\ c^2$ mit Spinozas Lehre hinweisen: "Die Körper sind mit Bezug auf Bewegung und Ruhe, Schnelligkeit und Langsamkeit, nicht aber in Bezug auf die Substanz, von einander unterschieden." (*Ethics* II, proposition 13, corollary 1). G. H. Shikmoni to O. Nathan, 14 January 1957. Einstein Archive, reel 33-311.

claim, that religious motivation corrupted the theory, was also made, though in a totally different context and for totally different intents. When, with the rise of Hitlerism in Germany, Einstein, the humanitarian, Jew, and pacifist, became the target of political and ideological attacks, his theory of relativity was proclaimed a typical product of "Jewish Physics," which tries to deprive true physics or "Aryan Physics" of its foundations. To substantiate this claim, Nazi ideologists tried to show that the development of the theory of relativity had been strongly influenced by the Talmud, that ancient body of religious and civil laws consisting of the Mishnah and Gemara, both commentaries on the Bible. The following excerpt from the *Zeitschrift für die gesamte Naturwissenschaft*, a periodical purposely founded for propaganda, is a typical example of such a Nazi polemic.

> The mode of thought that finds its expression in Einstein's theory is known, when applied to other ordinary things, as "Talmudic thinking." The task of the Talmud is to fulfill the precepts of the Torah, the Biblical law, by circumventing them. This is accomplished by means of suitable definitions of the concepts occurring in the law and by a purely formalistic mode of interpreting and applying them. Think about the Talmud Jew who places a food basket under his seat in a railway car, thus turning it formally into his residence and obeying thereby formally the law that on the Sabbath one should not travel more than a mile from his residence. It is this formal fulfillment that is important for the Jew. . . . This formalistic Talmudic thinking also manifests itself in Jewish physics. Within the theory of relativity, the principle of the constancy of the velocity

> of light and the principle of the general relativity of the phenomena in nature represent the "Torah," which under all circumstances must be fulfilled. For this fulfillment, an elaborate mathematical apparatus is necessary; and just as previously the concept of "residence" . . . was rendered lifeless and given a more expedient (zweckentsprechende) definition, so in the Jewish relativity theory, the notions of space and time are deprived of all spirit and defined in an expedient, purely intellectual way. This analogy is not accidental or artificial, it is deeply rooted in the very essense of Judaism. It exposes itself, for example, when Einstein declares in his "The Foundations of the General Theory of Relativity": "It will be seen from these reflections that in pursuing the general theory of relativity we shall be led to a theory of gravitation, since we are able to 'produce' a gravitational field merely by changing the system of coordinates."[83]

If we consider this statement from a purely factual point of view, that is, if we ignore its obviously anti-Semitic tone, it still remains a flagrant falsification for several reasons. First, there is no law, biblical or rabbinical, that allows traveling on Shabbat "in a railway car" or any other vehicle, except if it is a matter of preserving life, for the Shabbat was made for man and not man for Shabbat (Talmud, Yoma, 85 b). The Shabbat was ordained as a day of rest

[83] B. Thüring, "Physik und Astronomie in jüdischen Händen," *Zeitschrift für die gesamte Naturwissenschaft* 3, pp. 55–70 (May–June 1937); Bruno Thüring iterates here almost verbatim statements made by Hans Alfred Grunsky in his pamphlet *Der Einbruch des Judentums in die Philosophie* (Schriften der Deutschen Hochschule für Politik, Heft 14), (Junker & Dünnhaupt, Berlin, 1937), pp. 17–18.

even for slaves and animals to commemorate "creation and the redemption from Egypt" (Exodus 20:8; Deuteronomy 5:15).[84]

Second, it is absurd to claim that Einstein was influenced by the Talmud. Although he once declared "that as a child he received instruction in the Bible and in the Talmud," there can be no doubt that he never really studied the Talmud;[85] for German Jews, unlike the Jews of Eastern Europe, rarely read the Talmud. It might perhaps be objected that Einstein had been indirectly influenced by the Talmud through Spinoza, because the young Spinoza had studied the Talmud in Amsterdam when he was a disciple of Rabbi Manasse BenIsrael whose portrait has been immortalized by Rembrandt. It was not the Talmudist from Amsterdam, but the philosopher of Voorburg, the author of the *Ethics* whom Einstein admired and with whom he felt a kinship.

It is, of course, true that the Talmud, representing different rabbinical interpretations of biblical laws, has a distinct style of its own, not very dissimilar to that of the scholastic theological writings. Its often extremely meticulous distinctions were not intended, as Thüring and Grunsky assert, to make it possible to deceive oneself or somebody else, but merely to clarify unresolved obscurities. Finally, if it were true that the study of the Talmud, often regarded as promoting critical thinking, had really inspired the creation of the theory of relativity, which has rightly been called "one

[84] For details of the rabbinical interpretation of Exodus 16:29 ("Let no man go out of his place on the seventh day"), see *The Mishnah* (Shabbat Eruvin, chap. 3-5, Seder Moad), e.g., in the English translation (P. Kehati, ed., Elineri Library, Jerusalem, 1990), pp. 33–87.

[85] He made this declaration in his interview with Viereck (Viereck, *Glimpses of the Great*, and D. Brian, *Einstein—A Life*, p. 186).

of the great triumphs of human thought,"[86] then modern science would be deeply indebted to the Talmud.

Not only Einstein's greatest success, his general theory, but also what he once called, as reported by George Gamow, "the biggest blunder of my life" has been claimed to have been religiously motivated.[87] To understand this claim, we must recall that soon after having obtained the field equations of general relativity, Einstein applied them to the universe as a whole. His paper "Kosmologische Betrachtungen zur allgemeinen Relativitätstheorie," published in 1917, initiated the modern study of relativistic cosmology and raised thereby the status of cosmology, which theretofore had been a jumble of speculations, to that of a respectable scientific discipline.[88]

Einstein thought that his first cosmological solution of the field equations was a failure and rejected it because it yielded a nonstatic (expanding) universe. He thus missed the chance of announcing the expansion of the universe as perhaps the most important prediction of his general theory. That the universe is, in fact, steadily expanding was revealed only in the late 1920s by Edwin Powell Hubble's observations at the Mount Wilson Observatory. In 1917, Einstein modified the field equations by introducing—

[86] Address of the President, Sir J. J. Thomson, at the Anniversary Meeting, *Proceedings of the Royal Society, London, A96* (1919): 311–322, quotation on p. 317.

[87] G. Gamow, *My Worldline* (Viking Press, New York, 1970), p. 44.

[88] A. Einstein, "Kosmologische Betrachtungen zur algemeinen Relativitätstheorie," *Sitzungsberichte der Preussischen Akademie der Wissenschaften 1917*, part 1, pp. 142–152; "Cosmological Considerations on the General Theory of Relativity," *The Collected Papers of Albert Einstein*, vol. 6 (Princeton University Press, Princeton, N.J., 1996), pp. 543–551.

without violating their covariance—an additional term, the so-called "cosmological constant λ," in order to obtain a static unchanging universe. It was this introduction of λ that Einstein called his "biggest blunder." It has been suggested that Einstein committed this "blunder" because he was influenced by Spinoza who, in his *Ethics* declared, "God is immutable or [which is the same thing] all his attributes are immutable," and "an extended thing [like space] (and a thinking thing) are God's attributes."[89] In accordance with Spinoza, Einstein interpreted the term "endure" in the verse "the Heavens endure from everlasting to everlasting" in the sense of an immutable existence.

[89] "Deum, sive omnia Dei attributa esse immutabilia," *Ethics*, col. 2 to proposition 20, p. 1; "rem extensam (et rem cogitantem) Dei attributa esse," ibid., col. 2 to proposition 14.

CHAPTER 2

Einstein's Philosophy of Religion

WHEREAS the preceding chapter presented a biographical account of Einstein's personal attitude toward religion, the present chapter offers a systematic study of his writings on the philosophy of religion. Einstein persistently abstained from using the term "theology." He did so because he realized that his approach to religion differs essentially from that of professional theologians and especially from those for whom "theology is in possession of the truth, philosophy is in quest of the truth," but also, and perhaps more important, because his own religious convictions leave no room for theology. However, if theology is defined as a methodically formulated attempt to understand religious doctrines in general, and the nature of God in particular, then, I believe, the term also would have been acceptable to Einstein.[1]

Although there can be no doubt that the problem of the relation between religion and science had already attracted Einstein's attention in the 1920s, if not earlier, his first explicit essay on this subject dates from the end of 1930, and most of his writings on religion date from 1930 to 1941. These were also the years in which his interest in the philosophy of physics was particularly keen. In 1930, he wrote a short review of Siegfried Weinberg's book *Erkenntnistheorie* (*Theory of Knowledge*); in 1933, he delivered the Herbert Spencer lecture "On the Nature of Theoretical Physics"; in 1936, he wrote his essay "Physik und Realität" ("Physics and Reality"). Strictly speaking, the Einstein-

[1] The term "theology" was used, probably for the first time, by Plato in his *Politeia* (*Republic* II, 379a), then by Aristotle (*Metaphysics* III, 4, 1000 a 9), and throughout the Middle Ages. Spinoza also used it, e.g., in the title of his *Tractatus theologico-politicus* (1670).

Podolsky-Rosen paper of 1935 also belongs to this category, because, through the work of John Stewart Bell, it led to what Abner Shimony has called "experimental metaphysics," and made it possible, in the words of Edward Purcell, "to settle a philosophical problem in the laboratory."[2]

Einstein's interest in writing about religion was prompted by two interviews. The first of these took place with J. Murphy and J. W. N. Sullivan early in 1930 at Einstein's apartment in Berlin, Haberlandstrasse 5.[3] Murphy began the conversation by telling Einstein that, at a 1929 New York meeting of scientists, sociologist Harry Elmer Barnes suggested "that the time has come for science to give a new definition of God," and that, in the course of the public discussion of this remark, "the general contention of the preachers was that the introduction of God into scientific discourse was quite out of place; for science has nothing to do with religion." Einstein commented that in his opinion "both attitudes [that of Barnes and that of the preachers] disclose a very superficial concept of science and also of religion."

Murphy then asked Einstein whether modern science can provide "spiritual help and inspiration which organized religion seems unable to give." Einstein replied,

> Speaking of the spirit that informs modern scientific investigations, I am of the opinion that all the finer speculations in the realm of science spring from a

[2] J. T. Cushing and E. McMullin, eds. *Philosophical Consequences of Quantum Theory* (University of Notre Dame Press, Notre Dame, Ind., 1989), p. 60.

[3] A. Einstein, "Science and God."

> deep religious feeling, and that without such feeling they would not be fruitful. I also believe that, this kind of religiousness, which makes itself felt today in scientific investigations, is the only creative religious activity of our time. The art of today can hardly be looked upon at all as expressive of our religious instincts.

To Sullivan's question about whether scientific achievements can really express religious feelings and whether religion arises from an attempt to find meaning in life or from the presence of suffering in the world, Einstein replied, "That does not seems to me to be a very high conception of religion. The really great religious men did not have that idea in their minds."

The conversation then veered to the question of whether scientific theories—such as those created by Einstein and his colleagues Planck, Heisenberg, and Schrödinger—can be turned into a philosophy that establishes practical ideals of life. Einstein declared that our moral judgments, our sense of beauty, and religious instincts are "tributary forms in helping the reasoning faculty toward its highest achievements. You are right in speaking of the moral foundations of science; but you cannot turn it around and speak of the scientific foundations of morality." For science, Einstein continued, cannot teach men to be moral and "every attempt to reduce ethics to scientific formulae must fail."

At the end of the interview, Murphy asked Einstein whether he agreed with Spengler's (at that time much discussed) prophecy that Western European civilization would decline and fall because it was no longer creative in the arts, or whether he believed that science had taken up the torch which art has let fall, and that with this torch it is lighting up the unplumbed distance?" Einstein replied that "modern science does supply the mind with an object for contemplative exaltation. Mankind must exalt itself. *Sursum corda* is always its cry. Every cultural striving, whether it be religious or scientific, touches the core of the inner psyche and aims at freedom from the Ego—not the individual Ego alone, but also the mass Ego of humanity."

The second interview was a dialogue with the Indian mystical poet and philosopher Rabindranath Tagore, the Nobel Laureate for Literature in 1913. Tagore visited Einstein at his summer house in Caputh near Potsdam on July 14, 1930, after delivering the Hibbert Lecture at Oxford on the topic "The Religion of Man."[4] It is therefore not surprising that it was now Einstein who posed the questions.

Tagore and Einstein had much in common. Like Einstein, Tagore was a lover of music. "How many times," he wrote, "the music and the glow of sunset have brought to our hearts the pulsation of the limitless world, have inspired inexpressible, great music that has no connection with our everyday sorrow and happiness, that is a chant of the whole universe going round the temple of the Lord! Why only music? Beauty of nature also produces the same effect."

Like Einstein, Tagore did not find religion in scriptures, tradition, and rituals. Tagore believed in a spiritual world that is not separate from this world but is its innermost truth. "With the breath we draw we must always feel this truth that we are living in God. Born in this great universe full of infinite mystery, we cannot accept our existence as a momentary outburst of chance, drifting on the current of

[4] R. Tagore, *The Religion of Man* (George Allen & Unwin, London, 1932).

matter towards an eternal nowhere!" The universe, Tagore said, is permeated by an idea that "reveals itself in an eternal symphony."

All these statements, though expressed in the language of a poet and visionary, are in accord with Einstein's ideas. Both believed that "the Divine is not isolated from the world." They disagreed on one important issue, however. Although Einstein believed in two opposing conceptions about the nature of the universe—(1) "the world as a unity dependent on humanity," and (2) "the world as a reality independent of the human factor," Tagore declared:

> The world is a human world—the scientific view of it is also that of the scientific man. . . . Science is concerned with that which is not confined to individuals; it is the impersonal human world of truths. Religion realizes these truths and links them up with our deeper needs; our individual consciousness of truth gains universal significance. Religion applies values to truth, and we know truth as good through our own harmony with it.

On Tagore's comment that truth, like beauty, is not independent of human beings, Einstein asked: If there were no longer any human beings, would the Apollo of Belvedere no longer be beautiful? Tagore replied, "No." "I agree with regard to this conception of Beauty, but not with regard to Truth," replied Einstein. "Why, not? Truth is realized through man," declared Tagore. Einstein objected: "I cannot prove that my conception is right, but that is my religion. . . . If there is a *reality* independent of man, there is also a truth relative to this reality; and in the same way, the negation of the first engenders a negation of the latter."

At the end of the conversation, Tagore proclaimed, "If

there be some truth which has no sensuous or rational relation to the human mind, it will ever remain as nothing so long as we remain human beings," whereupon Einstein replied, "Then I am more religious than you are!"[5]

EINSTEIN, IT seems, was somewhat disappointed about the talk he had with Tagore. To make his views quite clear, he wrote an essay, "What I Believe," in the early fall of 1930.[6] An abridged version is his "Credo" (Glaubensbekenntnis), which was recorded phonographically in the fall of 1932 in Berlin on the intiative of the "Deutsche Liga für Menschenrechte" ("German League for Human Rights"); but the records were confiscated and destroyed by the Nazis. An excerpt published in the Berlin daily *Tempo* of January 24, 1933, was reprinted in David Reichinstein's biography of Einstein.[7] Fortunately, historian of science Friedrich Herneck succeeded in retrieving the original text and published it in 1966 in *Naturwissenschaften*.[8]

[5] "Note on the Nature of Reality," Appendix II in R. Tagore, *Religion of Man*, pp. 222–225; also in *Modern Review* (Calcutta) 49 (1931): 42–43; *American Hebrew* 129 (September 11, 1931): 351–355; *The Living Age* 340 (1931): 262–265.

[6] A. Einstein, "What I Believe," *Forum and Century* 84 (1930): 193–194; also in *Ideas and Opinions*, pp. 8–11; E. H. Cotton, ed., *Has Science Discovered God?* (Crowell, New York, 1931), pp. 93–97. Original German text in A. Einstein, *Mein Weltbild* (Querido, Amsterdam, 1934), pp. 11–17. *The World as I See It* (Lane, London, 1935), pp. 1–5.

[7] D. Reichinstein, *Albert Einstein, sein Lebensbild und seine Weltanschauung* (Stella, Prague, 1934); *Albert Einstein, a Picture of His Life and His Conception of the World* (E. Goldstone, London, 1934), pp. 99–100.

[8] F. Herneck, "Albert Einsteins gesprochenes Glaubensbekenntnis," *Naturwissenschaften* 53 (1966): 198; *Einstein und sein Weltbild* (Der Morgen, Berlin, 1976), pp. 99–101.

In this essay, Einstein described his view concerning the purpose of life; the ideals of kindness, beauty, and truth without which his life would be empty; his political ideals of democratic principles; and his hatred of any form of violence. The following two excerpts define his ethical and religious point of view.

> I do not at all believe in human freedom in the philosophical sense. Everybody acts not only under external compulsion but also in accordance with inner necessity. Schopenhauer's saying, "A man can do what he wants, but not want what he wants," has been a real inspiration to me since my youth; it has been a continual consolation in the face of life's hardships, my own and others', and an unfailing wellspring of tolerance.

> The most beautiful experience we can have is the mysterious. It is the fundamental emotion which stands at the cradle of true art and true science. Whoever does not know it and can no longer wonder, no longer marvel, is as good as dead, and his eyes are dimmed. It was the experience of mystery—even if mixed with fear—that engendered religion. A knowledge of the existence of something we cannot penetrate, our perceptions of the profoundest reason and the most radiant beauty, which only in their most primitive forms are accessible to our minds—it is this knowledge and this emotion that constitute true religiosity; in this sense, and in this alone, I am a deeply religious man. I cannot conceive of a God who rewards and punishes his creatures, or has a will of the kind that we experience in ourselves.

Einstein concluded the essay by saying that he was satisfied with "the awareness and a glimpse of the marvelous structure of the existing world, together with the devoted striving to comprehend a portion, be it ever so tiny, of the Reason that manifests itself in nature." Because this "Reason" or "superior Intelligence," as he called it elsewhere, manifests itself, in accordance with the Spinozistic identity of *ordo idearum* and *ordo rerum*, through an unrestricted determinism and causality in nature, which also excludes the possibility of free will, the two cited excerpts are logically coherent.

Einstein's religion was based on a firm belief in determinism and in an uncompromising denial of any anthropomorphism and anthropopathism in the notion of God. Einstein could have regarded his radical anti-anthropomorphism as a logical consequence of the biblical anti-iconic Second Commandment of the Decalogue (Exodus 20:4; Deuteronomy 5:8). He greatly respected Maimonides (Moses ben Maimon), the foremost Jewish philosopher of the Middle Ages. He once said of Maimonides that he "exerted a crucial and fruitful influence on his contemporaries and on later generations."[9] Maimonides devoted a whole chapter of his *Guide for the Perplexed* to the problem of how to conceive of God and concluded that "our knowledge [of God] consists in knowing that we are unable to comprehend Him."[10] Spinoza also regarded God as beyond description

[9] A. Einstein, "Moses Maimonides," a talk delivered at the celebration of the 800th anniversary of Maimonides' birth, *New York Times*, 15 April 1935, p. 15. Also in A. Einstein, *Essays in Humanism* (Philosophical Library, New York, 1950, 1985), pp. 114–115.

[10] Maimonides, *The Guide for the Perplexed*, M. Friedländer, ed. (George Routledge, London; Dutton, New York, 1942), p. 85.

and imagination. In a letter to a friend, he explained it geometrically: "On your question whether I have a clear idea of God as I have of a triangle, I would answer in the affirmative; but on your question whether I have a clear image of God as I have of a triangle, I would answer in the negative. For of God no image can be made."[11]

Einstein's concept of God, not unlike Spinoza's concept of a triangle, does not admit any mental image; God can be conceived only through the "rationality or intelligibility of the world which lies behind all scientific work of a higher order." As he once explained to a Japanese scholar, a deep feeling and his belief in a superior mind that reveals itself in the world of experience represent his concept of God. "In common parlance this may be described as 'pantheistic' (Spinoza). Denominational traditions I can only consider historically and psychologically; they have no other significance for me."[12]

Nevertheless, the historical development of these "denominational traditions" aroused Einstein's interest, as shown in his essay "Religion and Science," which he wrote expressly for the *New York Times Magazine* in the late fall of 1930.[13] Starting with the assumption that everything mankind has done or thought was somehow related to "the satisfaction of deeply felt needs and the assuagement of

[11] "Ad quaestionen tuam, an de Deo tam claram, quam de triangulo habeam ideam, respondeo affirmando; si me vero interroges, utrum tam claram de Deo, quam de triangulo habeam imaginem, respondebo negando. Deum enim non est imaginari." B. Spinoza, Epistola LVI, in *Opera*, vol. 4 (C. Winter, Heidelberg), p. 261.

[12] A. Einstein, *Gelegentliches* (Soncino Gesellschaft, Berlin, 1929), p. 9.

[13] A. Einstein, "Religion and Science," *New York Times*, 9 November 1930, section 5, pp. 1–2 (abridged), with a commentary by R. L. D., p. 2.

pain," Einstein raised the question of what precisely have been the needs and feelings that led to religious thought and belief in the widest sense of the words. For primitive man, he answered, it was primarily "fear that evokes religious notions, fear of hunger, wild beasts, sickness, and death." Because human understanding of causal connections was still poorly developed, the human mind created imaginary beings on whose will life or death and the health or sickness of the individual or of the society as a whole were supposed to depend. In order to appease these beings toward human beings, they were offered supplications and sacrifices, the earliest forms of prayers and religious rituals. Einstein calls this first stage of the development of religion "the religion of fear."

Clearly, Einstein's theory of the origin of religion excludes the idea of a religion by revelation, according to which God makes his reality known to man by his actions, as for example by his apparition to Moses according to the Old Testament, or by certain events, as by the birth, life, and death of Jesus Christ according to the New Testament, or by words delivered by an angel as the Koran teaches. The idea that fear is the origin of religion was not Einstein's invention. Confining ourselves only to German theological literature, which Einstein could have consulted, we find in P. Wilhelm Schmidt's monumental treatise *Der Ursprung der Gottesidee* abstracts of the works of several anthropologists and students of primitive cultures who propounded this theory of the origin of religion.[14] It is, of

[14] P. W. Schmidt, *Der Ursprung der Gottesidee*, 7 vols. (Aschendorff, Münster, 1912–1926) (vol. 12 appeared in 1955). The most prominent proponents of the "fear theory," referred to by Schmidt, are, in chronological order, Geo Gerland, Aloys Borchert, Karl Beth, and Nathan Söderblom.

course, very unlikely that Einstein had read any of these works. As we know from Talmey's report, he did read Ludwig Büchner's *Kraft und Stoff*.[15] A chapter entitled "Die Gottesidee" ("The Idea of God") begins with a quotation from the Roman writer Petronius, "Primus in orbe Deos fecit timor" [At the very first it was fear that created the Gods].[16] When Einstein wrote his essay "Religion and Science," Petronius's statement may have been still in the back of his mind.

The second stage in the development of religion is, according to Einstein, "the social or moral conception of God," which arises from the "desire for guidance, love, and support." It is the God who rewards and punishes, who comforts in distress and preserves the souls of the dead. The Old and New Testaments are for Einstein an admirable illustration of the transition from the religion of fear to the gradual predominance of the religion of morality, which still adheres to an anthropomorphic conception of God.

The idea of such a God (or gods) is an old one that can be traced to ancient Greek philosophers like Xenophanes of Colophon, who said that "if oxen or lions could paint,

[15] See M. Talmey, *The Relativity Theory Simplified*.

[16] Petronius, *Satyricon Reliquiae* (Teubner, Stuttgart, 1995), p. 185. Quoted in L. Büchner, *Kraft und Stoff*, 13th ed. (Thomas, Leipzig, 1874), p. 209. In the English translation, *Force and Matter* (Twentieth Century, New York, 1894), the quotation is replaced by a citation from Gustave Naquet: "Whenever knowledge takes a step forward, God recedes a step backwards." In his book *Der Gottes-Begriff* (Thomas, Leipzig, 1874), p. 12, Büchner also gives this quotation from Petronius in the section beginning with the statement: "Was nun die früheste Entstehung des Gottesbegriffes angeht, so kann wohl kein ernstlicher Zweifel darüber bestehen, dass seine eigentlichen Väter Unwissenheit und Furcht gewesen sind" ("Concerning the earliest origin of the notion of God, there can be no serious doubt that its real genitors were ignorance and fear.")

they would present their gods in the form of oxen or lions." More than two millennia later, the German poet Friedrich von Schiller declared, "In seinen Göttern malt sich der Mensch" [Man paints himself in his gods], a statement that is also quoted in Büchner's *Kraft und Stoff*.

The third stage of religious experience, which Einstein calls "the cosmic religious feeling," is "very difficult to elucidate . . . to anyone who is entirely without it, especially as there is no anthropomorphic conception of God corresponding to it." Because this is the kind of religion that Einstein adopted, it is appropriate to quote in detail how he conceived it, although part of it is already known to us from his Credo.

> The individual feels the futility of human desires and aims and the sublimity and marvelous order which reveal themselves both in nature and in the world of thought. Individual existence impresses him as a sort of prison, and he wants to experience the universe as a single significant whole. The beginnings of cosmic religious feeling already appear at an early stage of development, for example, in many of the Psalms of David and in some of the Prophets. Buddhism, as we have learned especially from the wonderful writings of Schopenhauer, contains a much stronger element of this. The religious geniuses of all ages have been distinguished by this kind of religious feeling, which knows no dogma and no God conceived in man's image; so that there can be no church whose central teachings are based on it. Hence, it is precisely among the heretics of every age that we find men who were filled with this highest kind of religious feeling and

> were in many cases regarded by their contemporaries as atheists, sometimes also as saints. Looked at in this light, men like Democritus, Francis of Assisi, and Spinoza are closely akin to one another.

Einstein then turned to the question of how this cosmic religious feeling can be communicated from person to person and how it can be stirred up in the first place, if it lacks any theology and visual imageries. At this junction, religion and science meet. As Einstein declared, "In my view, it is the most important function of art and science to awaken this feeling and keep it alive in those who are receptive to it." Einstein obviously admits that there are people who are not receptive to it or have no access to science and art. Although he does not explicitly refer to these people, it would be probably correct to say that he would have fully agreed with Johann Wolfgang von Goethe's poem:

> Wer Wissenschaft und Kunst besitzt,
> der hat auch Religion;
> Wer jene beiden nicht besitzt,
> der habe Religion.
> [He who possesses science and art
> possesses religion as well;
> He who possesses neither of these
> had better have religion.][17]

provided the term "religion," when used for the first time denotes religious feeling, and when used for the second time, denotes "institutional religion" or "religious instruc-

[17] J. W. v. Goethe, "Gedichte," in *Goethes Werke,* vol. 1 (Wegner, Hamburg, 1948, 1952), p. 367.

tion." As discussed above, Einstein had no objection to religious education if properly performed.

Einstein continued his essay by pointing out that the relation between religion and science, if conceived in the sense of his cosmic religion, can no longer be that of an irreconcilable antagonism as this relation had often been regarded in the past. For

> the man who is thoroughly convinced of the universal operation of the law of causation cannot for a moment entertain the idea of a being who interferes in the course of events—provided, of course, that he takes the hypothesis of causality really seriously. He has no use for the religion of fear and equally little for social or moral religion. A God who rewards and punishes is inconceivable to him for the single reason that a man's actions are determined by necessity, external and internal, so that in God's eyes he cannot be responsible, any more than an inanimate object is responsible for the motion it undergoes.

Einstein then reaffirmed his belief that cosmic religious experience "is the strongest and noblest driving force behind scientific research," and that Kepler and Newton had been able to spend long years of solitary work to unravel the secrets of celestial mechanics only because they had been imbued with such a religious feeling.

The publication of Einstein's essay "Religion and Science," which appeared on the first page of the *New York Times Magazine* section on Sunday, November 9, 1930, was a startling event that attracted wide public attention. On the very same Sunday, some preachers in the New

York area referred to it in their sermons, some of them quite disapprovingly. It is therefore not surprising that the following Monday, the *New York Times* carried a lead article under the appeasing title of "The Religion of Research," which began with these words: "Certainly those who make researches into the very sources of existence should be profoundly religious, though it cannot be admitted that they are . . . the only deeply religious people. The religious sense may dwell in the soul of the child or of the primitive nomad looking out upon the stars."[18] In the same issue, the *Times* also published an article, "Dr. Ward Attacks Einstein Theories," in which Harry Fredrick Ward of the Union Theological Seminary at New York's James Memorial Chapel criticized Einstein's unrestricted determinism, which excludes free will and consequently also the commission of sin in contradiction to the apostle Paul's profoundly religious definition of sin. Dr. Ward protests against the "intellectual complacency of this generation" and the view that sin should be made obsolete for the benefit of a more ethical humanity, "in which man would be rid of his sense of fear and guilt" as a consequence of Einstein's concept of a cosmic religion.[19]

On the same page, the *Times* also published "Einstein's Faith Defended," a brief report of Rabbi Solomon B. Freehof's sermon at Chicago's Free Synagogue. In his sermon on "The Religious Philosophy of Albert Einstein," Rabbi Freehof declared: "The anti-religious view of the universe looks upon the world as a clearly understood machine in

[18] *New York Times*, 10 November 1930, p. 18.

[19] Ibid., p. 22.

which every 'riddle' is either solved or on the way to solution. To Einstein, the universe is essentially mysterious. He confronts it with awe and reverence."

A much stronger contrast of opinions existed between Rabbi Krass and Dr. Sheen. Nathan Krass, professor of homiletics at New York's Jewish Institute of Religion and rabbi at Temple Emanuel, stated point-blank: "The religion of Albert Einstein will not be approved by certain sectarians but it must and will be approved by the Jews." In contrast, Dr. Fulton John Sheen, professor at the Catholic University of America and an ordained priest, chided the *New York Times* for having "degraded itself" by publishing this "sheerest kind of stupidity and nonsense." He asked whether anyone would be willing to lay down his life for the Milky Way, and concluded, "There is only one fault with his cosmical religion: he put an extra letter in the word—the letter 's.'"[20]

While these discussions went on, Einstein and his wife Elsa were crossing the Atlantic on the SS *Belgenland* for Einstein's second visit to the United States, following an invitation to spend some time at the California Institute of Technology in Pasadena. Einstein realized the impact of his essay as soon as the ship docked in New York City; whenever newspaper or radio reporters interviewed him, at least one of their questions referred to his ideas about the relation between religion and science. On January 2, 1931, two days after his arrival at Caltech, Einstein received a telegram from Jacob Landau, the managing director of the Jewish Telegraphic Agency in New York, which read as

[20] Quoted after R. W. Clark, *Einstein—The Life and Times*, p. 517.

follows: [Landau] "ASKS FOR PERMISSION TO REPUBLISH IN BOOKFORM EINSTEIN'S RECENT ARTICLE ON RELIGION IN NEW YORK TIMES STOP ARTICLE AROUSED WIDE INTEREST AND WOULD BE VERY WORTH WHILE FURTHER DISTRIBUTION STOP NEW YORK TIMES HAS NO OBJECTION STOP THANKING YOU IN ADVANCE AND WITH BEST WISHES JACOB LANDAU." Einstein agreed under the condition that the *Times* truly had no objection.[21]

The record of reactions to Einstein's essay may suggest that the Christian clergy rejected, while the Jewish clergy approved, Einstein's cosmic religion, and that the latter were perhaps prejudiced because of Einstein's Jewish extraction. It should be noted, however, that the opinions recorded were only those expressed either by extremely orthodox Christian theologians or by extremely liberal Jewish religionists. As it happened, only such extremists publicized reactions to the 1930 essay. However, had we considered quite generally the relevant writings also of liberal Christians and of orthodox Jews, we would have arrived at a more balanced conclusion. To mention just two examples, the physicist Elmer L. Offenbacher, a former president of the Association of Jewish Orthodox Scientists, listed in an essay a number of reasons why "we are tempted to condemn his [Einstein's] concept of God,[22] and Hans Küng, a highly regarded Christian theologian of the University of Tübingen in Germany, wrote in his book *Does God Exist?* a rather sympathetic chapter on Einstein's reli-

[21] Einstein Archive, reel 47-097.

[22] E. L. Offenbacher, "Albert Einstein—Scientist and Jew," *Jewish Life*, May–June 1955: 15–23.

giosity, though of course not identifying himself with Einstein's philosophy of religion.[23]

ON JANUARY 4, 1931, before a congregation at Temple Isaiah in Chicago, Dr. Jacob Singer, rabbi at Temple Mizpah in Chicago, delivered a sermon that tried to represent both the orthodox and the liberal point of view. Singer began by quoting from Einstein's "Credo": "The most beautiful experience we can have is the mysterious. It is the fundamental emotion which stands at the cradle of true art and true science . . . it is this emotion that constitutes true religiosity; in this sense, and in this alone, I am a deeply religious man."[24] Singer, obviously referring again to the 1930 essay, continued:

> The numerous comments which this credo has aroused, ranging from the condemnation of the Fundamentalists to the laudations of the Liberals, indicate that religion is very much alive in our own day. Theologians have found ground for dissent with a "cosmic religion" which may become a Pantheism almost identical with Atheism. . . . When Einstein formulates his creed, he becomes a theologian, sharing with other theologians their lack of consistency and mental persuasiveness. We must realize however, that a man's religion is immeasurably larger than the theory about his faith.

Singer concluded his sermon with these words: "The magnitude of Einstein's soul eclipses the limitations of his the-

[23] H. Küng, *Does God Exist?* (Doubleday, Garden City, N.Y., 1980), pp. 627–666.

[24] See A. Einstein, "What I Believe."

ology. As a pioneer of the spirit, Einstein has scaled heights from which a new perspective of life has been disclosed."[25]

Einstein's insistence on an all-pervading unrestricted causal determinism was the main reason that, in general, the clergy, regardless of denomination, rejected the philosophy of cosmic religion. For the supreme reign of such a determinism denies not only the possibility of a divine interference—even if the existence of a personal God were admitted—it also deprives man of his free will and, as a consequence, of his moral responsibility. For if man's actions are dictated by this determinism, he cannot be held responsible and hence cannot be punished or rewarded for what he does.

Searching for a loophole in the logic of this argumentation to save the existence of moral values, Rabbi Abraham Geller of Brooklyn, New York, wrote a number of letters to Einstein in which he suggested, inter alia, relaxing the stringency of determinism by adopting the probabilistic theory of quantum mechanics. Geller apparently attached great importance to this idea, for he sent it, together with Einstein's reply, to the *New York Times*, which published it at the end of January 1931, in an article with the title "Einstein Explains Deterministic View" and with the subtitle "Brooklyn Scientist [sic] Reveals Letter in Answer to His Queries on Quantum Theory—Dr. Millikan Is Puzzled."[26]

In his reply, Einstein stated that he remained a convinced determinist because he had no doubt that the "theoretical basis [of quantum mechanics] will be replaced later by a deterministic" theory and that it also be noted "that

[25] J. Singer, *Einstein's Religion* (a sermon published as a pamphlet in 1931).

[26] *New York Times*, 28 January 1931, p. 2, col. 2.

the deterministic conception of life-events is not substantially influenced by the other theory." Robert Andrews Millikan, who knew Einstein personally, remarked: "Least of all, in my mode of thinking, is it possible for a man to be a real determinist who, like Einstein, has a sense of 'social responsibility,' for the sense of social responsibility means freedom of choice and self-condemnation as a result of wrong choices." In a long letter to Einstein, Geller apologized for having divulged this correspondence without permission. Einstein obviously accepted the apology for he continued to correspond with Geller. In his last letter to Geller in 1933, Einstein reaffirmed his view that the volitions of man are part of the necessary course of events and causally bound just as the revolutions of the planets; but the theoretical endorsement of determinism in no way affects the demands of practical ethics; it only leads to a tender-hearted and conciliatory (milderen und verständnisvollen) disposition to other people and to a mitigation of the feeling of hatred and enmity. "Only the idea of a God who punishes and rewards is not compatible (except in an artificial way) with the thesis of determinism."[27]

Einstein wrote this letter in the Belgian resort Le Coq-sur-Mer where he terminated his originally planned return journey from the United States to Germany, for, as he told a friend, he "dared not enter Germany because of Hitler" who four weeks earlier, on January 30, 1933, had seized power. In Le Coq-sur-Mer, Einstein for the first time reconsidered his pacifistic call to refuse military service in any war whatever its motivations. But he continued to adhere to his Spinozistic deterministic belief all through his life.

[27] Einstein to Geller, 17 April 1933. Einstein Archive, reel 33-293.

As late as 1948, answering Besso's commendation of the Christian maxim "Love thy enemy," Einstein wrote that he agreed as far as actions are concerned. "But for me," he continued, "the cogitative basis is the trust in an unrestricted causality. 'I cannot hate him, because he *must* do what he does.' That means, for me more Spinoza than the Prophets."[28]

And yet, after the horrors of the concentration camps had become known, Einstein declared, "The Germans as an entire people are responsible for these mass murders and must be punished as a people if there is justice in the world."[29] He never forgave the Germans for having voted for Hitler, and he hated the perpetrators of those horrible crimes against humanity. He may have been aware that he acted, strictly speaking, against his conviction in an unrestricted determinism and against Spinoza's teaching that "Hatred can never be good" [Odium nunquam potest esse bonum, *Ethics*, part 4, proposition 45]. Einstein could, however, justify himself on the grounds that, as he once said, despite his deterministic constraints, man must conduct his moral life as if he were free. Or, as he wrote in his 1930 essay "Religion and Science," "the ethical conduct of man should be based effectually on sympathy, education, and social ties and needs; no religious basis is necessary."[30]

[28] "Deine Bemerkung über Feindes-Liebe teile ich, was die Einstellung zum Handeln anlangt. Bei mir ist aber die gedankliche Basis das Vertrauen auf die unbeschränkte Kausalität. 'Ich kann ihn nicht hassen, weil er so thun *muss*, wie er thut.' Also bei mir mehr Spinoza als die Propheten." Einstein to M. Besso, 6 January 1948. *Albert Einstein—Michele Besso, Correspondance 1903–1955* (Hermann, Paris, 1972), p. 392. Einstein Archive, reel 7-382.

[29] A. Einstein, *Out of My Later Years*, p. 239.

[30] A. Einstein, *Ideas and Opinions*, p. 39.

In a short article, published in 1934, he had emphasized that "there is nothing divine about morality; it is a purely human affair." The main issue dealt with in this article was the distinction between the religion of "the naive man," for whom God is a being who rewards or punishes, and the religiosity of the scientist that consists of "a rapturous amazement at the harmony" of the laws of nature. The scientist's religious feeling "is the guiding principle of his life and work, insofar as he succeeds in keeping himself from the shackles of selfish desire. It is beyond question closely akin to that which has possessed the religious geniuses of all ages."[31]

Nowhere in his entire 1930 essay nor anywhere else in his writings did Einstein discuss the concept of a miracle, although miracles play an important role in almost every religion. Miracles center chiefly around Moses during the Exodus from Egypt in biblical Judaism, around Jesus as healer in Christianity, and around Mohammed and his ascent to heaven in Islam. The only record we have of Einstein's concept of miracles is a report of a conversation he had in 1931 with David Reichinstein, whose name has been mentioned in connection with Einstein's "Credo."

A miracle is usually defined as an event that violates the laws of nature or as the intrusion of the supernatural into the realm of the natural. It is therefore clear that Einstein's belief in an unrestricted determinism in science, where an unalterable antecedent-consequent relation is a necessary condition for the comprehensibility of experience (essentially a Kantian idea), is incompatible with a belief in the existence of

[31] A. Einstein, *The World as I See It*, p. 28; *Ideas and Opinions*, p. 40.

miracles. But as we learn from his conversation with Reichinstein, for Einstein, the belief in determinism was not a necessary condition for the denial of miracles.

Reichinstein told Einstein about the thesis of Arthur Liebert, a professor of philosophy and editor of the *Kant-Studien*, that Heisenberg's indeterminacy relations of quantum mechanics, or more generally the indeterminism of that theory, imply that "the possibility of a miracle needs no longer to be excluded." Perhaps recalling how a few weeks earlier at the Sixth Solvay Congress in Brussels, Niels Bohr had successfully rebutted his attempt to disprove the Heisenberg relations,[32] Einstein did not challenge the use and validity of these relations. Instead, he replied that he could not accept Liebert's argument because it dealt "with a domain in which lawful rationality [determinism] does not exist. A 'miracle,' however, is an exception from lawfulness; hence, there where lawfulness does not exist, also its exception, i.e., a miracle, cannot exist."[33]

It should be noted that Einstein's argument does not disprove the possibility of what may be regarded as a miraculous event; it only denies the legitimacy of regarding such an event as a miracle. This is not the place to mention recent discussions on the relation between quantum mechanics and religion. Suffice it to point out that all miracles recorded in the history of religion were macroscopic processes for which the statistical laws of quantum me-

[32] For details on the Bohr-Einstein debate, see M. Jammer, *The Philosophy of Quantum Mechanics* (Wiley, New York, 1974), chap. 5.

[33] "Dort, wo eine Gesetzmässigkeit nicht vorhanden ist, kann auch ihre Ausnahme, d.h. ein Wunder, nicht existieren." D. Reichinstein, *Die Religion der Gebildeten* (Verlag Aristoteles, Zurich, 1941), p. 21.

chanics, generally speaking, agree with the laws of deterministic physics in accordance with the so-called Ehrenfest's theorems.

When the Northwestern Regional Conference of the American Association of Theological Schools convened at the Theological Seminary in Princeton in May 1939, one of the few nontheologians invited to address the meeting was Einstein. The mimeographed transcripts of his lecture carried the title "The Goal."[34] Einstein began his talk by recalling that in the last century it was widely held that scientific knowledge and religious belief conflict with each other and that the prevailing trend "among advanced minds" was to replace belief with knowledge. The function of education was therefore confined to the development of rational thinking and knowing. Although "the aspiration toward such objective knowledge belongs to the highest of which man is capable . . . knowledge of what *is* does not open the door directly to what *should* be. One can have the clearest and most complete knowledge of what *is*, and yet not be able to deduct from that what should be the *goal* of our human aspirations." Scientific thinking alone, Einstein continued, cannot lead us to the ultimate and fundamental purpose of our existence.

> To make clear these fundamental ends and valuations, and to set them fast in the emotional life of the individual, seems to me precisely the most important function which religion has to perform in the social life of man. And if one asks whence derives the authority of such fundamental ends, since they cannot be

[34] A. Einstein, "The Goal," lecture delivered 19 May 1939, *Ideas and Opinions*, pp. 41–44; *Out of My Later Years*, pp. 25–28.

> stated and justified merely by reason, one can only answer: they exist in a healthy society as powerful traditions, which act upon the conduct and aspirations and judgments of the individuals; they are there, that is, as something living, without its being necessary to find justification for their existence. They come into being not through demonstration but through revelation, through the medium of powerful personalities. One must not attempt to justify them, but rather to sense their nature simply and clearly. The highest principles for our aspirations and judgments are given to us in the Jewish-Christian religious tradition. It is a very high goal, which, with our weak powers, we can reach only very inadequately, but which gives us a sure foundation to our aspirations and valuations.

Compared with his 1930 essay, this talk had a much more reserved tone and its ideas were acceptable even to orthodox theologians. It should be noted, however, that the topic of Einstein's 1930 essay differs distinctly from that of his 1939 talk; while the former dealt mainly with the origin and nature of religious beliefs, the latter deals almost exclusively with questions related to the purpose and goal of our life, a subject on which agreement is more easily attainable than on the nature of religion. In fact, Einstein's 1939 talk was sympathetically received by almost all participants of the conference.

This was probably one of the reasons that Rabbi Louis Finkelstein, a prominent religious leader, president of the Jewish Theological Seminary in New York, and member of the organizing committee of the "Conference on Science, Philosophy and Religion," scheduled to convene on September 9–11, 1940, at the Union Theological Seminary in the City of New York, thought it appropriate to invite Einstein to address this conference as well. Einstein agreed to write an essay, "Science and Religion," to be read at this conference.[35] Neither he nor Finkelstein anticipated the serious controversies and harsh acrimonies that this essay would evoke.

Einstein agreed, not only out of respect for a distinguished leader of liberal Judaism but also because of his well-known magnanimity to respond to all requests he thought to be ingenuous. Thus, in 1936 when Phyllis Wright, a sixth-grade student in the Sunday school of the Riverside Church in New York, asked whether scientists pray and, if they do, what they pray for, he gave a reply that can serve as an introduction to his essay for the 1940 conference.

> Scientific research is based on the assumption that all events, including the actions of mankind, are determined by the laws of nature. Therefore, a research scientist will hardly be inclined to believe that events could be influenced by a prayer, that is, by a wish addressed to a supernatural Being. However, we have to admit that our actual knowledge of these laws is only an incomplete piece of work (unvollkommenes Stückwerk), so that ultimately the belief in the existence of fundamental all-embracing laws also rests on a sort of faith. All the same, this faith has been largely justified

[35] A. Einstein, "Science and Religion," *Transactions of the First Conference on Science, Philosophy and Religion in Their Relation to the Democratic Way of Life* (New York, 1941); *Ideas and Opinions*, pp. 44–49; *Out of My Later Years*, pp. 28–33; *Nature* 146 (1940): 605–607.

by the success of science. On the other hand, however, every one who is seriously engaged in the pursuit of science becomes convinced that the laws of nature manifest the existence of a spirit vastly superior to that of men, and one in the face of which we with our modest powers must feel humble. The pursuit of science leads therefore to a religious feeling of a special kind, which differs essentially from the religiosity of more naive people. With friendly greetings, your Albert Einstein."[36]

EINSTEIN'S CONTRIBUTION to the 1940 conference was presented to an audience of over five hundred participants. The article begins with the question of what, precisely, we understand by science and by religion. Science, says Einstein, can easily be defined as "the attempt at the posterior reconstruction of existence by the process of conceptualization"; but to define religion is a much more difficult task. We can reach this definition by inquiring first what characterizes the aspirations of a religious person. "A person who is religiously enlightened," says Einstein, "appears to me to be one who has, to the best of his ability, liberated himself from the fetters of his selfish desires and is preoccupied with thoughts, feelings, and aspirations to which he clings because of their superpersonal value." What is important, according to Einstein, is "the force of this superpersonal content, . . . regardless of whether any attempt is made to unite this content with a divine Being." From these presuppositions, Einstein then derived the definition

[36] Einstein to P. Wright, 24 January 1936. Einstein Archive, reel 52-337.

of religion as "the age-old endeavor of mankind to become clearly and completely conscious of these values and goals and constantly to strengthen and extend their effect."

These definitions enabled Einstein to repeat what he had already said in his essay, "The Goal," namely, that because science ascertains only what *is*, but not what *should be*, no conflict between the two can exist. Only intervention on the part of religion into the realm of science—if, for example, a religious community insists on the absolute truthfulness of all statements in the Bible—can give rise to conflict, as has been the case in the struggle of the Church against the doctrines of Galileo or Darwin. Even though the realms of religion and science are distinctly marked off from each other, strong reciprocal relations exist between the two. Though religion determines the goal, science, in its broadest sense, shows the means for attaining this goal. However, "science can only be created by those who are thoroughly imbued with the aspiration toward truth and understanding. This source of feeling, however, springs from the sphere of religion. . . . I cannot conceive of a genuine scientist without that profound faith. The situation may be expressed by an image: science without religion is lame, religion without science is blind."

Had this statement been the final conclusion, the article probably would have been acclaimed by all the participants. But Einstein qualified his statements about the compatibility of religion and science "with reference to the actual content of historical religions." "This qualification," he continued, "has to do with the concept of God." He then mentioned, though more briefly than in his 1930 essay, his theory of the three stages in the evolution of religion and the concept of God and declared that "the main source of the present-day conflicts between the spheres of religion and of science lies in this concept of a personal God." Although he conceded that the doctrine of a personal God could never be refuted, because such a doctrine could always take refuge where science has not yet been able to gain a foothold, he called such a procedure

> not only unworthy but also fatal. For a doctrine which is able to maintain itself not in clear light but only in the dark, will of necessity lose its effect on mankind, with incalculable harm to human progress. In their struggle for the ethical good, teachers of religion must have the stature to give up that source of fear and hope which in the past placed such vast power in the hands of priests. The further the spiritual evolution of mankind advances, the more certain it seems to me that the path to genuine religiosity does not lie through the fear of life, and the fear of death, and blind faith, but through striving after rational knowledge. In this sense I believe that the priest must become a teacher if he wishes to do justice to his lofty educational mission.

Some background is necessary to assess correctly the reaction that this article—in particular, its denial of a personal God—evoked among the theologians attending the conference and the wider public. Einstein did not anticipate that the denial of a personal God would be misinterpreted as the denial of God. That such a misinterpretation was not uncommon can be gathered from a 1945 encyclopedia of religion that defined the term "atheism" as "the denial that there exists a being corresponding to some particular definition of god; frequently, but unfortunately, used to denote the denial of God as personal."[37] That Einstein was neither an atheist nor an agnostic—certainly not in the usual sense of the term coined in 1869 by Thomas Henry Huxley—follows not only from Einstein's abovementioned statements concerning his cosmic religion but also from statements made by all those with whom he had intimate discussions about his religious conviction. Thus, for example, his close friend Max Born once remarked, "he [Einstein] had no belief in the Church, but did not think that religious faith was a sign of stupidity, nor unbelief a sign of intelligence."[38] David Ben-Gurion—who visited Einstein in Princeton a year before inviting him to become President of Israel—recalled that, when discussing religion, "even he [Einstein], with his great formula about energy and mass, agreed that there must be something behind the energy."[39] With respect to religion, Ben-Gurion and Einstein had much in common. Like Einstein, Ben-Gurion was an ardent admirer of Spinoza. He also declared his belief "that there must be a being, intangible, indefinable, even unimaginable, but something infinitely superior to all we know and are capable of conceiving,"[40] a belief not much different from Einstein's belief in the impersonal God of his cosmic religion.

At a charity dinner in New York, Einstein explicitly dissociated himself from atheism when he spoke with the German anti-Nazi diplomat and author Hubertus zu Löwen-

[37] V. Ferm, ed., *An Encyclopedia of Religion* (Philosophical Library, New York, 1945), p. 44.

[38] *Born–Einstein Letters* p. 203.

[39] M. Pearlman, *Ben Gurion Looks Back* (Weidenfeld and Nicolson, London, 1965), p. 217.

[40] Ibid., p. 216.

stein: "In view of such harmony in the cosmos which I, with my limited human mind, am able to recognize, there are yet people who say there is no God. But what really makes me angry is that they quote me for support of such views."[41]

As shown by the numerous criticisms of his 1940 article and the many letters he received from people who read about it in the press, Einstein was approvingly quoted "for support of such views" by freethinkers, agnostics, and atheists, just as he was strongly reproached by orthodox extremists and fundamentalists. Einstein described the reaction to his article quite caustically.

> I was barked at by numerous dogs who are earning their food guarding ignorance and superstition for the benefit of those who profit from it. Then there are the fanatical atheists whose intolerance is of the same kind as the intolerance of the religious fanatics and comes from the same source. They are like slaves who are still feeling the weight of their chains which they have thrown off after hard struggle. They are creatures who—in their grudge against the traditional "opium for the people"—cannot bear the music of the spheres. The Wonder of nature does not become smaller because one cannot measure it by the standards of human moral and human aims.[42]

Only a small fraction of the numerous responses to Einstein's 1940 essay will be discussed, for an exhaustive account of all the articles in the newspapers and of all the

[41] Prinz Hubertus zu Löwenstein, *Towards the Further Shore* (Victor Gollancz, London, 1968), p. 156.

[42] Einstein to an unidentified addressee, 7 August 1941. Einstein Archive, reel 54-927.

letters that Einstein received in reaction to his essay would require at least a separate chapter. Generally speaking, the newspapers tried to be noncommittal or to present the reactions of both opponents and supporters alike. Thus, for example, the *New York Times*, of 11 September 1940, published a rather brief note under the heading "Religion of God Urged by Einstein—He Tells Philosophers It Is More Worthy than 'Concept of a Personal God.'" After reporting that Einstein "decries fear as a basis" of religion, the article declared that "various other plans were offered for unifying democracy against totalitarianism," obviously avoiding a controversy about religious issues. In contrast, one day later, the *Hudson (N.Y.) Dispatch* published a series of commentaries on Einstein's essay written mostly by clergymen of different denominations but all united in their disapproval of Einstein's denial of a personal God. Thus, the Reverend Carl F. Weldmann, pastor of the Lutheran Church of Our Saviour, Woodcliff (N.J.) is reported to have declared: "God, who has indicated his power in the order and beauty of the world, and who has given us the ability to say 'I' and who has granted us a personality, must of necessity be a person himself." The Reverend Joseph Antliff of St. Brigid's Roman Catholic Church, North Bergen, New Jersey, said, "I have not read the story about Einstein, but the conception of any God but a personal one is contradictory to the teachings of the Catholic Church." He was seconded by a priest of a North Hudson (N.Y.) Catholic church who withheld his name: "There is no other God but a personal God. . . . Einstein does not know what he is talking about. He is all wrong. Some men think that because they have achieved a high degree of learning in some field, they are qualified to express opinions in all."

Rabbi Hyman Cohen of the West New York and Guttenberg Talmud Torah commented: "Einstein is emphatically no atheist. He believes in a God. But in renouncing a personal deity he removes the Supreme Being so remotely from the sphere of human comprehension as to make His influence on the individual's conduct negligible. To use mathematical terminology, he reduces the infinite to an infinitesimal of the highest order. . . . Einstein is unquestionably a great scientist, but his religious views are diametrically opposed to Judaism."

The *Kansas City Times* cited Bishop Edwin Vincent O'Hara of the Catholic diocese of Kansas City as saying, "It is sad to see a man, who comes from the race of the Old Testament and its teaching, deny the great tradition of that race." But according to Rabbi Gershon Hadas of Beth-Sholom Synagogue in Kansas City, the issue is only a matter of difference in the terminology of our expression about faith in God. "Herein lies the confusion. If we translate Einstein's terms into the formal language of religious expression I have a feeling that Einstein's cry in a wilderness of worlds and aeons is the self-same cry that comes from the heart of all religious folk." The Kansas daily also reported that Rabbi Samuel S. Mayerberg saw "nothing sensational in Dr. Einstein's statements" and that Methodist lay leader Leslie J. Lyons "cautioned that the same words could be variously interpreted by many people, particularly in religious affairs. . . . I personally still retain complete faith in the doctrine of a personal God." Dr. Burriss Jenkins of the Community Church "discerned nothing dangerous in the Einstein point of view. . . . It's an old concept at that, as old, say, as the Hindu religion. Many see no personal God, but I am sure the results in personal conduct are little different from those reaped when the old theological conception of a God of punishment and wrath prevailed." Mrs. Renick Martin, chairperson of the Kansas women's division of the Christian mission, sees Einstein as "trying to understand God in the light of his experience as a scientist and great scholar, just as we in our small way try to understand God as housewives and mothers."

The leading article in the *Detroit Free Press* of September 14, 1940, written by its editor Malcolm Walker Bingay, an Episcopalian, used much tougher language. He wrote that Einstein's advice

> to give up the doctrine of a personal God . . . shows that the good Doctor, when it comes to the practicalities of life, is full of jellybeans. . . . The conditions of the world being what they are, he does his own people a grave injury by making public such a statement. By so doing, he is giving the religious bigots, especially the followers of Hitler and the Ku Klux Klan, fuel for their fanatical fires. They will charge that he is presenting the Jewish faith when, as a matter of fact, what he is saying is an utter denial of the whole Jewish concept of God.

Many more commentaries, generally less depreciatory than Mr. Bingay's, could be cited from the *(New York) Herald Tribune*, the *New York Evening Sun*, and other dailies, or from *Time Magazine*, which declared that "Einstein's message was the only false note of the entire conference." All of these responses were rather vague or emotional and did not really touch on the core of Einstein's argumentation.

A more detailed criticism was published on September 21, 1940, by the Catholic periodical the *Tablet* under the heading "Monsignor Sheen Answers Einstein's Anti-God Attack." Sheen, who had severely criticized Einstein's 1930 essay, was asked to reply to the 1940 essay. As he wrote in the *Tablet*, he initially declined to comment because "just as I would not expect Professor Einstein to make a reply to anything I might write on relativity, so Professor Einstein would not expect me to make a reply to anything he might say about religion, for both of us would be talking about something we know nothing about." Sheen recalled that he had refused a request to write "an article of rejoicing" after Eddington and Jeans announced in an international broadcast that experimental physics had "discovered" God. He felt that their reasons for the existence of God were "just as flimsy and false as the reasons Professor Einstein gave for the denial of a personal God." Just as "the shoemaker should stick to his last," as the ancient proverb says, so also said Sheen, the "scientist should stick to his science," and he who talks religion should at least say his prayers. If "the thinking processes of Professor Einstein were not keener in the field of science than they are in the field of philosophy," Sheen declared, "then he has made me a skeptic of relativity."

Turning to the refutation of Einstein's ideas, Sheen pointed out Einstein's denial of a personal God was based on two arguments. The first argument contended that the concept of a personal God is "a sublimation of the old conception of the gods." But, said Sheen, Einstein gave "not a spark of evidence" that the belief in many gods preceded a belief in God nor did he show how a "sublimation" produces a personal God. If this were true, then Einstein's cosmic religion would be a sublimation of Protagoras's individual relativity. Furthermore, Sheen contended, if most people never heard of primitive religions, how could their belief in a personal God be a sublimation of that primal mythology? Einstein's second argument, continued Sheen, rested on the irreconcilability of the notion of a personal omnipotent God and the idea of the moral responsibility of man: God would not be all-powerful if He made man free. According to Sheen,

> it seems much more in accordance with sound reason to argue that God would not be all-powerful unless He could make man free. It takes more skill to make a machine that runs itself than it does to make a bird house. In like manner, it takes more power to make a self-determining human being than it does to make an automaton. An omnipotent God, Professor Einstein says, would make man irresponsible. It is just the contrary, for how can there be responsibility without personality. The moral order assumes law, and law is based on Mind, and Mind is personal. If God is only impersonal Space-Time, there is no moral order; then Hitler is not responsible for driving Professor Einstein out of Germany. It was only a bad collocation of space-time conglomerations which made him act that way.

Sheen concluded his commentary by agreeing with Einstein that the future of mankind depends on "a striving after rational knowledge" and also that his [Einstein's] arguments would not win the audience to his belief. Addressing Einstein directly, Sheen declared, "That is true, Professor. But the reason will not be because you are not the world's greatest scientist. You are. It will be because when you stepped out of your laboratory, you failed to use 'rational knowledge.'"

To assess the public's reception of Einstein's denial of a personal God, excerpts from some of the many letters that Einstein received on the issue are quoted here. The authors of these letters are identified only by their initials.

A totally disabled veteran of World War I and, as he called himself, a patriotic citizen of the United States of America, wrote from Rochester, New York: "The great leaders, thinkers and patriots of the past who fought and died for free thought, free speech, free press, and intellectual liberty arise to salute you! With the great and mighty Spinoza, your name will live as long as humanity."[43]

A member of the Society of Jesus and of the faculty of Georgetown University in Washington, D.C., sent the following letter:

> I should like to commend your great wisdom in recently presenting the problem concerning the cooperation of God with the operations of men. May I suggest that you will find an adequate treatment of the subject in the first book of Thomas Aquinas' *Summa contra Gentiles*, and also in the first part, question III, article 8 of the *Summa Theologica*, which St. John's College of Annapolis has placed on its list of the world's one hundred best books together with your own great work on relativity.[44]

[43] AGC to Einstein, 11 September 1940. Einstein Archive, reel 40-247.

[44] JFG, S.J. to Einstein, 11 September 1940. Einstein Archive, reel 40-367. JFG apparently referred to *Summa contra Gentiles*, chap. 7: "quod veritati fedei christianae contraritur veritas rationis," chap. 20: "quod deus non est corpus," and chap. 44: "quod deus est intelligens."

A historian and president of a historical society in New Jersey had this to say to Einstein:

> We respect your learning, Dr. Einstein; but there is one thing you do not seem to have learned—That God is a spirit and cannot be found through the telescope or microscope, no more than human thought or emotion can be found by analyzing the brain. As everyone knows, religion is based on Faith, not knowledge. Every thinking person, perhaps, is assailed at times with religious doubt. My own faith has wavered many a time. But I never told anyone of my spiritual abberations for two reasons: (1) I feared that I might, by mere suggestion, disturb and damage the life and hopes of some fellow being; (2) because I agree with the writer who said, "There is a mean streak in anyone who will destroy another's faith." . . . I hope, Dr. Einstein, that you were misquoted and that you will yet say something more pleasing to the vast number of the American people who delight to do you honor.[45]

A Catholic attorney-at-law and active member of an organization of Catholics, Jews, and Protestants to promote better understanding between all creeds and classes informed Einstein that

> we deeply regret that you made your statement before the Conference on Science, Philosophy and Religion in which you ridicule the idea of a personal God. In the past ten years nothing has been so calculated to make people think that Hitler had some reason to expel the

[45] HWE to Einstein, 14 November 1940. Einstein Archive, reel 40-339.

> Jews from Germany as your statement. Conceding your right to free speech, I still say that your statement constitutes you as one of the greatest sources of discord in America.[46]

Another lawyer, from Fredericksburg, Virginia, who likewise introduced himself as having "endeavored to stem the tide of growing anti-Semitism in the United States," wrote:

> Unfortunately, and perhaps unconsciously, with the sole exception of German propaganda, the strongest provocative force with which non-Jewish friends of the Jews have to contend emanates from the Jews themselves. I say "unfortunately and unconsciously" because I sincerely believe that you are among the number who are adding fuel to the fire, and believe me, Doctor Einstein, fuel is being added to the fire, and there is unquestionably a growing spirit of anti-Semitism in the United States, and, very unfortunately, I think it is growing in the South where it has never been before. You did a great disservice to your people in your recent pronouncement about the eminence of a personal Providence. Whether you believe it or not, as a reputed distinguished scholar, why should you rush to the Press to give your views on the most provocative of all subjects, that is, Religion? In fact, shortly after your pronouncement, I heard a great deal of criticisms of it with the remark that Einstein was a *brainy* Jew, but that Religion was not in the sphere of the brains but in the heart, and that there have been many

[46] RRC to Einstein, 19 September 1940. Einstein Archive, reel 40-330.

> poor Jewish women around the Wailing Wall in Jerusalem who had more Faith than Einstein, and could teach him much.[47]

Probably the most strongly worded letter came from the founder of the Calvary Tabernacle Association in Oklahoma City, Oklahoma, who wrote:

> Professor Einstein, I believe that every Christian in America will answer you, "We will not give up our belief in our God and His Son Jesus Christ, but we invite you, if you do not believe in the God of the people of this Nation, to go back where you came from." I have done everything in my power to be a blessing to Israel, and then you come along and with one statement from your blasphemous tongue do more to hurt the cause of your people than all of the efforts of the Christians who love Israel can do to stamp out anti-Semitism in our Land. Professor Einstein, every Christian in America will immediately reply to you, "Take your crazy, fallacious theory of evolution and go back to Germany where you came from, or stop trying to break down the faith of a people who gave you a welcome when you were forced to flee your native land."[48]

In addition to articles in the press and personal letters, a book, *The Fifth Column in Our Schools*, condemned Einstein's denial of a personal God. On October 1, 1940, a few weeks before the conference at the Union Theological Seminary took place, Einstein, his stepdaughter Margot, and his secretary Helen Dukas were sworn in as U.S. citi-

[47] ATE to Einstein, 3 October 1940. Einstein Archive, reel 40-343.
[48] EFW to Einstein, 12 September 1940. Einstein Archive, reel 40-372.

zens in Trenton, New Jersey, by Judge Philipp Forman. Concerning Einstein's declaration "so help me God" at the conclusion of his oath of allegiance, *The Fifth Column* stated,

> If Albert Einstein is right and there is no personal God, then America is founded on fable and falsehood. . . . If there is no God, then the citizen has no God-given rights. Then all the rights set forth in the constitution are sham and delusion. If man has no Creator, then our fathers fought for a lie; then the rights of citizenship are based on a lie. Then Professor Einstein has subscribed to a lie in the very act of pledging allegiance to a form of government which, according to his philosophy, is founded on a lie.[49]

AMONG THE critics of Einstein's denial of a personal God was Paul Johann Oskar Tillich. Unlike most commentators, Tillich wrote a rather dispassionate, almost sympathetic, commentary, trying to offer "a solution in which [Einstein's argument] is accepted and overcome at the same time."

Tillich was the first non-Jewish professor in Germany to be exiled by the Hitler regime because of his close affiliation with the German Social-Democratic party and his opposition to Nazi ideology. In 1933, he emigrated to the United States, where he became one of the most prominent philosophical theologians. As his influential *Systematic Theology* shows, he had given much thought to the notion of a personal God long before he heard Einstein's address.[50]

[49] Quoted after R. W. Clark, *Einstein—The Life and Times*, pp. 686–687.

[50] P. Tillich, *Systematic Theology*, vol. 1 (University of Chicago Press, Chicago, 1951; Nisbet, London, 1953), chap. 8–10.

Now he thought it appropriate to subject Einstein's argumentation to a profound philosophical, almost logical, analysis. He began his critical essay by stating that "Einstein attacks the idea of a personal God from four angles: the idea is not essential for religion; it is the creation of primitive superstition; it is self-contradictory; it contradicts the scientific world view."[51]

Because the first argument implies the concept of religion, which is defined as "the acceptance of and devotion to superpersonal values," and because the question of whether this definition is adequate cannot be answered before "the question is answered whether the idea of the Personal God has some objective meaning or not . . . we must turn to the second argument, the historical." This argument, Tillich continued, does not, and cannot, show "why primitive imagination created just the idea of *God*." True, "mythological phantasy can create stories about Gods but it cannot create the idea of God itself, because the idea transcends all elements of experience which constitute mythology." Having refuted the second argument and therefore, by implication, the cogency of the first argument, Tillich turned to the third argument which claims that the notion of a personal, omnipotent, and righteous God is self-contradictory because omnipotence and the existence of moral and physical evil are incompatible. Yet, said Tillich, this argument

[51] P. Tillich, "The Idea of a Personal God," *Union Review* 2 (1940): 8–10; "Das Problem des 'persönlichen Gottes,'" in P. Tillich, *Gesammelte Werke*, vol. 12 (Evangelisches Verlagswerk, Stuttgart, 1971), pp. 300–304. For a recent discussion of this problem, without mentioning Einstein's address, see W. J. Mander, "God and Personality," *Heythrop Journal* 38 (1997): 401–412.

> presupposes a concept of omnipotence which identifies omnipotence with omni-activity in terms of physical causality. But it is an old and always emphasized theological doctrine that God acts in all beings according to their special nature, in all beings according to their rational nature, in animals and plants according to their organic nature, in stones according to their inorganic nature. The symbol of omnipotence expresses the religious experience that no structure of reality and no event in nature and history has the power of preventing us from community with the infinite and inexhaustible ground of meaning and being.

Before Tillich turned to the last and, what he rightly called the "most important" of Einstein's four arguments, he made two methodological remarks. Tillich fully agreed with Einstein's rejection of the use of any *argumentum ex ignorantia*; he agreed "entirely with Einstein when he warns the theologians not to build their doctrines in the dark spots of scientific research" because any interference of theology with the task of science is destructive to theology itself.

Second he asked Einstein "to deal with theology in the same fairness which is demanded from everyone who deals, for instance, with physics—namely to attack the most advanced and not obsolete forms of a discipline." One should not use a primitive pattern of the concept of the personal God in order to challenge the idea itself. In order to show that a mature idea of the personal God need not, and in fact cannot, interfere with science or philosophy, Tillich quoted Einstein's own words of "that humble attitude of mind toward the grandeur of reason incarnate in existence, which, in its profoundest depths, is inaccessible to man." Tillich interpreted these words as pointing "to a common ground of the whole of the physical world and of the superpersonal values, a ground which, on the one hand, is manifest in the structure of being, (the physical world) and meaning (the good, true, and beautiful)—which, on the other hand, is hidden in its unexhaustible depth." This, according to Tillich, is the fundamental element of every developed conception of God from the earliest Greek philosophers to present-day theology. Because of its cardinal importance for the whole issue under discussion, Tillich's conclusion is quoted verbatim in spite of its length.

> The manifestation of this ground and abyss of being and meaning creates what modern theology calls "the experience of the numinous." Such an experience can occur in connection with the intuition of the "grandeur of reason incarnate in existence," it can occur in connection with the belief in "the significance and loftiness of those superpersonal objects and goals which neither require nor are capable of rational foundation"—as Einstein says. The same experience can occur—and occurs for the large majority of men—in connection with the impression some persons, historical or natural events, objects, words, pictures, tunes, dreams, etc. make on the human soul, creating the feeling of the holy, that is of the presence of the "numinous." In such experiences, religion lives and tries to maintain the presence of and community with this divine depth of our existence. But since it is "inaccessible" for any objectivating concept, it must be ex-

> pressed in symbols. One of these *symbols* is the "Personal God." It is the common opinion of classical theology, practically in all periods of Church history, that the predicate "personal" can be said of the Divine *only symbolically or by analogy* or if affirmed and negated at the same time. It is obvious that, in the daily life of religion, the symbolic character of the idea of the "Personal God" is not always realized. This is dangerous only if distorting theoretical or practical consequences are derived from the failure to realize it. Then attacks from outside and criticism from inside follow and must follow. They are demanded by religion itself. Without an element of "atheism," no "theism" can be maintained.[52]

At the very end of his commentary, Tillich raised the question of whether the symbol of the personal needs to be used; and he answered this question by referring to a term used by Einstein himself, the "suprapersonal." For, quoting Tillich again,

> the depth of being cannot be symbolized by objects taken from a realm which is lower than the personal, from the realm of things or subpersonal living beings. The suprapersonal is not an "It," or more exactly, it is a "He" as much as it is an "It," and it is above both of them. But if the "He" element is left out, the "It" element transforms the alleged suprapersonal into a subpersonal, as it usually happens in monism and pantheism. And such a neutral subpersonal cannot grasp the center of our personality; it can satisfy our aesthetic

[52] Emphases added.

> feeling or our intellectual needs, but it cannot convert our will, it cannot overcome our loneliness, anxiety, and despair. . . . This is the reason that the symbol of the personal God is indispensable for living religion. It is a symbol, not an object, and it never should be interpreted as an object.

Tillich had communicated with Einstein before 1940, though not on religious or philosophical problems but on the project of a self-taxation of German immigrants for the benefit of fellow exiles in need ("Selbsthilfe Deutscher Ausgewanderter").[53] It is not known whether Einstein received a reprint of Tillich's article in the *Union Review* or how he reacted to the publication, in particular, Tillich's statement that the notion of a personal God is only a symbol, though a necessary one. Tillich had already published this idea before Einstein wrote his 1940 essay.[54] An expanded version, with the same title, appeared three years after Einstein's death.[55] It would greatly enhance our understanding of Einstein's philosophy of religion to know how he would have responded to these articles, because Tillich's statements converge toward Einstein's "cosmic religion" as much as is possible for a theistic theologian. The following quotation from Tillich gives some evidence of the similarity of their positions.

> In the word "God" is contained at the same time that which actually functions as a representation and also

[53] See, e.g., Tillich to Einstein, 21 October 1937. Einstein Archive, reel 54-353.

[54] P. Tillich, "The Religious Symbol," *Journal of Liberal Religion* 2 (1940): 13–33.

[55] P. Tillich, "The Religious Symbol," *Daedalus (Proceedings of the American Academy of Arts and Sciences)* 87 (1958): 3–21.

> the idea that it is *only* a representation. It has the peculiarity of transcending its own conceptual content: upon this depends the numinous character that the word has in science and in life in spite of every misuse through false objectification. God as an object is a representation of the reality ultimately referred to in the religious act, but in the word "God" this objectivity is negated and at the same time its representative character is asserted.[56]

Tillich was a Protestant theologian, but the eminent Catholic theologian Hans Küng—who was affiliated with the University of Tübingen as Tillich had been before his emigration—expressed similar ideas.

> If Einstein speaks of cosmic reason, this must be understood as an expression of reverence before the mystery of the Absolute, as opposed to all-too-human "theistic" ideas of God. . . . God is not a person as man is a person. The all- embracing and all-penetrating is never an object that man can view from a distance in order to make statements about it. . . . [God] is not an individual person among other persons, is not a superman or superego. The term "person" also is merely a cipher for God.

What Tillich has called "symbol," Küng calls "cipher," denoting not zero but rather "code" or "cryptogram."[57]

It should be pointed out that when theologians like Tillich or Küng call the personal God a "symbol," they

[56] Cf. also P. Tillich, *Dynamics of Faith* (Harper, New York, 1957); *Gesammelte Werke*, vol. 8, pp. 111–196, esp. chap. 3 (Die Symbole des Glaubens), pp. 139–148.

[57] H. Küng, *Does God Exist?* p. 632.

mean not only a "sign," but also a religious symbol designed to reveal a dimension of reality that nonsymbolic language is incapable of expressing. It is a reality that transcends human comprehension and control but inspires men with a feeling of the sacred, or, in Rudolf Otto's terminology, of the numinous, which evokes an awe-inspiring *mysterium tremendum et fascinens*.[58] This emotion is precisely the "awe" or "cosmic religious feeling" that Einstein repeatedly declared to be the source of his religiosity and that, in his words, "is very difficult to elucidate . . . to anyone who is entirely without it, especially as there is no anthropomorphic conception of God corresponding to it."[59] The religious philosophy of Einstein and that of these leading theologians are, after all, not as disparate as may have been expected.

ONLY EIGHT years after his address to the Union Theological Seminary Einstein again agreed to write an address on a religious theme. Early in 1948, Alson Robinson, a neighbor and friend of Einstein, and Jacob Trapp of New York brought Einstein greetings from the Liberal Ministers' Club of New York City, an organization comprising mainly Unitarian clergymen. Seventeen prominent members of the club signed the letter, including John Howland Lathrop, president of the International Association for Liberal Christianity and Religious Freedom, and Frederick Robertson Griffin, director of the American Unitarian Association. Einstein was deeply moved by this expression of respect. In response, he agreed to write an essay entitled "Religion

[58] R. Otto, *The Idea of the Holy* (Oxford University Press, New York, 1950), p. 12; *Das Heilige* (Beck, Munich, 1917).

[59] See, e.g., Einstein's 1930 essay "Religion and Science."

and Science: Irreconcilable?" which was to be read at the forthcoming meeting of the club and to be published in its official organ, *The Christian Unitarian Register*.[60] It dealt with two questions: "Does there truly exist an insuperable contradiction between religion and science?" and "Can religion be superseded by science?" Einstein declared that the answer to both questions is negative. To prove this claim he argued as follows. At first he defined science as "systematical thinking directed toward finding regulative connections between our sensual experiences" [planmässiges Denken, darauf gerichtet, ordnende Beziehungen zwischen den sinnlichen Erlebnissen der Menschen zu finden].[61] Science can lead to systematic action provided definite goals are set up in advance. Although science is capable of examining whether such goals and valuations ("Ziele und Wertungen") are compatible or incompatible, these goals and values cannot be defined by science because they belong to the domain of religion. Religion, Einstein continued, is "concerned with man's attitude toward nature at large, with the establishing of ideals for the individual and communal life, and with mutual human relationship." [Zur Domäne der Religion gehört die Einstellung des Menschen zur Natur im Grossen, die Aufstellung von Idealen über die Ziele der Individuen und der Gemeinschaft und über das gegenseitige Verhalten der Menschen.] Religion attempts to attain these ideals

> by exerting an educational influence on tradition and through the development and promulgation of certain

[60] A. Einstein, "Religion and Science: Irreconcilable?" *Christian Unitarian Register* 127 (June 1948): 19–20; *Ideas and Opinions*, pp. 49–52.

[61] German original text in Einstein Archive, reel 29-118.

easily accessible thoughts and narratives (epics and myths) which are apt to influence evaluation and action along the lines of the accepted ideals. It is this mythical, or rather this symbolic, content of the religious traditions which is likely to come into conflict with science. This occurs whenever this religious stock of ideas contains dogmatically fixed statements on subjects which belong in the domain of science. Thus, it is of vital importance for the preservation of true religion that such conflicts be avoided when they arise from subjects which, in fact, are not really essential for the pursuance of the religious aims.

Einstein declared that, in his view, if the existing religions are divested of their myths, they do not basically differ from each other because the moral attitudes of a people, supported by religion, must always aim at "preserving and promoting the sanity and vitality of the community." To honor "falsehood, defamation, fraud, and murder" would lead to communal disintegration. If we look at the present state of civilized humanity from the standpoint of even the most fundamental religious commands, we must feel deeply and painfully disappointed.

For while religion prescribes brotherly love in the relations among the individuals and groups, the actual spectacle more resembles a battlefield than an orchestra. Everywhere, in economic as well as in political life, the guiding principle is one of ruthless striving for success at the expense of one's fellow men. This competitive spirit prevails even in school and, destroying all feelings of human fraternity and cooperation, conceives of achievement not as derived from the love for productive and thoughtful work, but as springing from personal ambition and fear of rejection.

To show that this deplorable situation is not a necessity of nature Einstein referred to certain so-called primitive cultures, like that of the Pueblo Indians of Arizona or New Mexico,[62] who lived under the hardest living conditions and still "accomplished the difficult task of delivering its people from the scourge of competitive spirit and of fostering in it a temperate, cooperative conduct of life, free of external pressure and without any curtailment of happiness."

Einstein concluded his essay with the following remarks.

> The interpretation of religion, as here advanced, implies a dependence of science on the religious attitude, a relation which, in our predominantly materialistic age, is only too easily overlooked. While it is true that scientific results are entirely independent from religious or moral considerations, those individuals to whom we owe the great creative achievements of science were all of them imbued with the truly religious conviction that this universe of ours is something perfect and susceptible to the rational striving for knowledge. If this conviction had not been a strongly emotional one and if those searching for knowledge had not been inspired by Spinoza's *Amor Dei Intellectualis*, they would hardly have been capable of that untiring devotion which alone enables man to attain his greatest achievements.

[62] In this context, Einstein mentioned Ruth Benedict's book *Patterns of Culture* (Houghton Mifflin, Boston, 1934), the fourth chapter of which describes the social conditions of the pueblos of New Mexico.

Unlike the 1940 essay, this essay seems to have been received favorably almost without exception, although probably not all of its statements have been met with approval by every member of the Liberal Ministers' Club. Such a favorable reception was not surprising because, compared with the 1940 essay, this essay was, almost a sermon in content and style. Here Einstein applied the term "religion" in the traditional sense and not in the sense of "cosmic religion." Einstein alluded to the conception of "cosmic religion" only at the very end of the essay, where he said that "the truly religious conviction that this universe of ours is . . . susceptible to the rational striving for knowledge" and where he mentioned Spinoza's *Amor Dei Intellectualis*.

Einstein's last message on a religious issue was addressed to a public forum that was even more liberal than the Liberal Ministers' Club, namely the Ethical Culture Society. Founded in 1876 by Felix Adler, who began his career with a study for the rabbinate and ultimately became a professor of ethics at Columbia University, this society chose as its motto "Deed Rather than Creed." It was founded on the maxim that "the true test of religious consecration is not in the worship of the supernatural but in what men do for one another on this earth to achieve mutually creative and liberating relationships."

When the Ethical Culture Society celebrated its seventy-fifth anniversary on January 5, 1951, President Harry Truman and Einstein were among those who paid tribute to the society, hailing it as "a stabilizing moral influence at a time of world crisis."[63] In his message, "The Need for Ethi-

[63] *New York Times* 6 January 1951, p. 16, col. 6.

cal Culture," Einstein acknowledged that the society had accomplished much in its efforts to combat prejudice and superstition. The removal of obstacles, he warned, does not by itself ennoble the standard of individual and social life.

> For along with this negative result, a positive aspiration and effort for an ethical-moral configuration of our common life is of overriding importance. Here no science can save us. I believe, indeed, that overemphasis on the purely intellectual attitude, often directed solely to the practical and factual, in our education, has led directly to the impairment of ethical values. I am not thinking so much of the dangers with which technical progress has directly confronted mankind, as of the stifling of mutual human considerations by the "matter-of-fact" habit of thought which has come to lie like a killing frost upon human relations.[64]

THE STATEMENTS and the general tenor of Einstein's last two essays can give the impression that, in the last years of his life, Einstein recanted his cosmic religion and sympathized with the traditional conception of religious faith, influenced perhaps by Schopenhauer's dictum that "religion could not have arisen had it not been for the fact of death." These essays may also suggest that Einstein abandoned, in particular, his thesis that religion and ethics are independent of each other, for how could he otherwise declare, for example, that "religion prescribes brotherly love"? Einstein's correspondence from the last years of his life shows that he did not change his conception of religion. That he

[64] A. Einstein, "The Need for Ethical Culture," in *Mein Weltbild*; *Ideas and Opinions*, pp. 53–54.

also persisted in the conceptual separation between religion and ethics follows, for example, from his correspondence with his old friend Maurice Solovine. At the end of 1950, Solovine translated Einstein's *Out of My Later Years* into French[65] and criticized some passages from the book. He was particularly intrigued by Einstein's statement that "science without religion is lame, religion without science is blind."[66] He asked, "If a serious difficulty is encountered in any branch of science, can religion provide any assistance? And conversely, if in religion any difficulty arises concerning the exegesis of a dogma or the removal of an inconsistency, can science provide any help?"[67]

Einstein replied that he understood Solovine's "aversion to the use of the word 'religion' when what is meant is an emotional or psychological attitude, which is most obvious in Spinoza." But he added,

> I have found no better expression than "religious" for confidence in the rational nature of reality as it is accessible to human reason. Wherever this feeling is absent, science degenerates into uninspired empiricism. . . . I cannot accept your opinion concerning science and ethics or the determination of aims. What we call science has the sole purpose of determining what *is*. The determining of what *ought to be* is unrelated to it and cannot be accomplished methodically. Science can only arrange ethical propositions logically and furnish the means for the realization of ethical aims, but

[65] A. Einstein, *Conceptions scientifiques, morales et sociales* (Flammarion, Paris, 1952).

[66] See A. Einstein, "Science and Religion."

[67] M. Solovine to Einstein, 7 December 1950. Einstein Archive, reel 21-273.

> the determination of aims is beyond its scope. At least that is the way I see it.[68]

Many more excerpts from Einstein's correspondence or conversations in the early fifties are evidence that he remained faithful to his religious conviction until his death in April 1955.

In December 1952, Mrs. Beatrice F. of San Francisco, California, wrote to Einstein that she has had "a rigid orthodoxly religious" education but has later "turned more and more to atheism and agnosticism," always thinking that Einstein, too, was "a free thinker, possibly atheist or agnostic." However, when reading Lincoln Barnett's book, *The Universe and Dr. Einstein*, she continued, "I see that the author says 'Dr. Einstein believes that a Divine Mind is the God of the Universe.' I would very, very much appreciate it," she concluded her letter, "if you would kindly clarify this for me—what your belief is in regard to Divinity. It means a great deal to me to know. Very sincerely yours, (Mrs.) Beatrice F."[69]

Einstein's reply is a very concise summary of his Credo.

> The idea of a personal God is quite alien to me and seems even naive. However, I am also not a "Freethinker" in the usual sense of the word because I find that this is in the main an attitude nourished exclusively by an opposition against naive superstition. My feeling is insofar religious as I am imbued with the consciousness of the insufficiency of the human mind

[68] Einstein to M. Solovine, 1 January 1951. Einstein Archive, reel 21-274; reprinted in A. Einstein, *Letters to Solovine* (Philosophical Library, New York, 1987), pp. 119–121.

[69] BF to Einstein, 12 December 1952. Einstein Archive, reel 59-794.

> to understand deeply the harmony of the Universe which we try to formulate as "laws of nature." It is this consciousness and humility I miss in the Freethinker mentality. Sincerely yours, Albert Einstein.[70]

In the summer of 1954, less than a year before he died, Einstein was interviewed by Professor William Hermanns. A German veteran of World War I, Hermanns began his career as a diplomat but, because of his opposition to the Nazi regime, he emigrated to the United States where he taught German literature at Harvard University and San José State University and then worked at the Hoover Institution at Stanford University. Hermanns had interviewed Einstein several times before, the first time in 1930 in Berlin. He fully endorsed Einstein's ideas of a cosmic religion. In fact, a survivor of the gruesome battle of Verdun, he intended to promote peace by founding a cosmic-religion movement including the Jewish, Christian, Vedic, Buddhist, and Islamic traditions; he claimed that Einstein's cosmic religion does not destroy the religious values of traditional beliefs but rather embraces them in accordance with the Biblical word, "Hear O Israel, the Eternal is our God, the Eternal is one." It was therefore only natural that the discussion turned to the notion of cosmic religion in the 1954 interview. Hermanns asked Einstein for "precise statements on God." Einstein replied,

> About God, I cannot accept any concept based on the authority of the Church. As long as I can remember, I have resented mass indoctrination. I do not believe in the fear of life, in the fear of death, in blind faith. I

[70] Einstein to BF, 17 December 1952. Einstein Archive, reel 59-797.

> cannot prove to you that there is no personal God, but if I were to speak of him, I would be a liar. I do not believe in the God of theology who rewards good and punishes evil. My God created laws that take care of that. His universe is not ruled by wishful thinking, but by immutable laws.[71]

These statements show clearly that Einstein never renounced the tenets of his cosmic religion. Is it really true that Einstein never spoke of a "personal God"? Do not his words "My God created laws" ascribe some "personality" to God, not necessarily, of course, in the anthropomorphic sense of the term? Einstein used a similar expression many years earlier in a talk with Esther Salaman, a student of physics, when he said, "I want to know how God created this world. I'm not interested in this or that phenomenon, in the spectrum of this or that element. I want to know His thoughts, the rest are details."[72]

It seems legitimate to ask whether an attribution of thoughts to God does not imply the notion of a personal God. That, conversely, the concept of a personal God entails the attribution of thought can hardly be contested even if it is conceded that the thoughts of God are not the thoughts of men or, as Isaiah phrased it, "My thoughts are not your thoughts, neither are your ways my ways."[73] But the God of the Bible is, of course, a personal God. Furthermore, many philosophers—among them, the pre-Socratic Alcmaion of Crotos—characterize a personal being or man

[71] W. Hermanns, *Einstein and the Poet—In Search of the Cosmic Man* (Branden Press, Brookline Village, Mass., 1983), p. 132.

[72] E. Salaman, "A Talk with Einstein," *The Listener* 54 (1955): 370–371.

[73] Isaiah 55:8. See also Paul's Epistle to the Romans in which God's thoughts are said to be "inscrutable." Romans 11:33.

by having the faculty of thinking apart from the faculty of perceiving.[74] It may therefore be claimed that when he was speaking about the thoughts of God, Einstein was speaking about a personal God in contradiction to his repeated declarations that such an idea is "quite alien" to him. The charge of being inconsistent would lose its validity only if it could be proved that he had used this expression in merely a metaphorical sense.

To examine the validity of this theory, let us quote a similar statement that Einstein made to Ernst Gabor Straus, his assistant from 1944 to 1948: "What I am really interested in is knowing whether God could have created the world in a different way; in other words, whether the requirement of logical simplicity admits a margin of freedom."[75] The second part of this statement, beginning with "in other words" indicates that Einstein's reference to God in the first part was merely a manner of speaking.

As mentioned above[76] and as Einstein himself has repeatedly emphasized—for example, in his Credo, where he said that "the most beautiful experience we can have is the mysterious" and "it is this emotion that constitutes true religiosity"[77]—the notion of "awe" or "cosmic religious

[74] See, e.g., H. Diels, *Die Fragmente der Vorsokratiker* (Weidmannsche Verlagsbuchhandlung, Berlin, 1951), p. 215; K. Freeman, *Ancilla to the Pre-Socratic Philosophers* (B. Blackwell, Oxford, 1956), p. 40.

[75] "Was mich eigentlich interessiert, ist, ob Gott die Welt hätte anders machen können; das heisst, ob die Forderung der logischen Einfachheit überhaupt eine Freiheit lässt." E. Straus, "Assistent bei Albert Einstein," in C. Seelig, *Helle Zeit—Dunkle Zeit* (Europa Verlag, Zurich, 1956), p. 72.

[76] A. Einstein, *Conceptions scientifiques.*

[77] F. Herneck, *Naturwissenschaften* 53 (1966): 198.

feeling," which carries a connotation of the mysterious, lies at the foundation of his philosophy of religion. Therefore, it is reasonable to ask, Was Einstein a mystic? That a brilliant physicist or mathematician can also be a devoted mystic, though perhaps not necessarily of the ecstatic Dionysian type, is well illustrated by the case of Blaise Pascal.[78] A positive answer to this question seems to be suggested by Einstein's lifelong enchantment with music and, as is often said, "nowhere does mysticism find more complete expression than in music."[79] A positive answer is also common among those who understand little of the theory of relativity but know that it considers time a fourth dimension, which they interpret as a vindication of spiritualism. The following episode provides an example.

When Einstein and his wife Elsa visited London in 1921, they attended a private dinner party at Queen Anne's Gate. In the course of their conversation, Randall Thomas Davidson, the Archbishop of Canterbury, said to Einstein that Viscount Haldane of Cloan, Einstein's host, had once told him that the theory of relativity "ought to make a great difference to our morale." Einstein replied, "Do not believe a word of it. It makes no difference. It is purely abstract—science." Later Davidson's wife discussed this matter with Elsa and told her of a friend who was enthusiastic about Einstein's theory "especially about its mystical aspect." Elsa "broke into laughter with the words: 'Mystical! Mystical! My husband mystical!'" echoing his own reply to a Dutch woman he met in the German embassy in

[78] B. Pascal, *Pensées sur la Religion* (Wolfgang, Amsterdam, 1688).
[79] V. Ferm, *Encyclopedia of Religion*, p. 514.

The Hague who said that she liked his mysticism. "Mysticism is in fact the only reproach that people cannot level at my theory," he had replied.[80]

Einstein himself expressed his personal aversion to mysticism several times. When he was asked, early in 1921, whether he believed that the soul exists and continues to exist after death, he replied, "The mystical trend of our present time, showing itself especially in the exuberant growth of the so-called Theosophy and Spiritualism is for me only a symptom of weakness and confusion."[81] In 1955, he wrote, "What I see in nature is a magnificent structure that we can comprehend only very imperfectly, and that must fill a thinking person with a feeling of 'humility.' This is a genuinely religious feeling that has nothing to do with mysticism."[82]

This statement probably provides the correct answer to the question of whether Einstein was a mystic. First, what he previously called a "feeling of "awe" he called here "a feeling of humility," and "humility" has nothing to do with mysticism. Furthermore, if mysticism denotes immediate intuition of, or insight into, a spiritual truth in a way different from ordinary sense perception or the use of logical thinking, Einstein was never a mystic. He never main-

[80] Quoted after R. W. Clark, pp. 339–340.

[81] "Der mystische Zug unserer Zeit . . . ist für mich nur ein Symptom von Schwäche und Zerfahrenheit." Einstein to Lili Halpern-Neuda, 5 February 1921. Einstein Archive, reel 43-847.

[82] "Was ich in der Natur sehe, ist eine grossartige Struktur, die wir nur sehr unvollkommen zu erfassen vermögen und die einen vernünftigen Menschen mit einem Gefühl der 'Humility' erfüllen muss. Dies ist ein echt religiöses Gefühl, das nichts mit Mystizismus zu schaffen hat." H. Dukas and B. Hoffmann, *Albert Einstein—The Human Side* (Princeton University Press, Princeton, N.J., 1979), p. 132.

tained that any knowledge—of the holy or the profane—could be attained through extrasensory perception or spiritual insight. He never conceived of his "cosmic religious feeling" as a substitute for rational thinking. When he said that "the most beautiful and deepest experience a man can have is the sense of the mysterious," he immediately added that "it is the underlying principle of religion as well as of all serious endeavor . . . in science." In short, in spite of his use of words like "awe," his philosophy of religion never transcended the realm of the rational.

THE ABSENCE of any mystic or other irrational components in Einstein's philosophy of religion makes it possible to regard it as a theoretical construction or system of ideas that can be subjected to methodological criteria for its acceptability, such as having only well-defined concepts or freedom of self-contradiction. Although some of its theses, such as the denial of a personal God, have been criticized, no critique of the whole system and its logical foundations seems to have been made before 1950. Solovine's letter cited above may mark the beginning of such critical assessments.[83]

Solovine challenged Einstein's conception of cosmic religion, which is the core of Einstein's philosophy of religion, on the grounds that its use of the term "religion" differs from ordinary usage of the word. Moreover, Solovine argued, Einstein's ascribing to religion the assignment and attainment [die Setzung und Erreichung hoher Ziele] of higher ideals presupposes "the existence of institutions and people to carry out this task," in contradiction to Ein-

[83] See M. Solovine to Einstein, 7 December 1950.

stein's rejection of institutional religions. Furthermore, Solovine claimed, religion does not inspire the progress of science, as Einstein tried to prove by invoking the examples of Kepler and Newton; rather, the progress of science "purifies" (läutert) and "humanizes" (macht menschlich) religion. Einstein's reaction to Solovine's criticisms has already been quoted above.[84]

Although Solovine criticized Einstein only for having misinterpreted the relation between science and religion in the case of Kepler, Benedictine priest and science historian Stanley L. Jaki charged Einstein with having contradicted himself. Jaki accused Einstein of having been "surely inconsistent" when stating in his Introduction to a biography of Kepler that "Kepler was a pious Protestant" but also that "his scientific success depended on his 'freeing himself to a large extent from the spiritual tradition in which he was born.'"[85] Moreover, according to Jaki the fact "that Einstein was not taken to task by historians of science for speaking of Kepler as a freethinker, illustrate[s] the negative attitude toward Christian theism which prevails in influential scholarly circles nowadays."[86]

Another question is how far, if at all, the basic ideas of Einstein's cosmic religion had been anticipated by philosophers or theologians of the nineteenth and the early twentieth centuries. Tillich began his essay with the statement that "if it had not been Einstein, the great transformer of

[84] See Einstein to Solovine, 1 January 1951.

[85] A. Einstein, Introduction to C. Baumgardt, *Johannes Kepler* (V. Gollancz, London, 1952), pp. 12–13.

[86] S. L. Jaki, *The Road of Science and the Ways to God* (Chicago University Press, Chicago, 1978), pp. 349–350.

our physical world view, his arguments would have not produced any excitement. They are neither new nor powerful in themselves. But in the mouth of Einstein, as an expression of his intellectual and moral character, they are more significant than the highly sophisticated reasoning of somebody else."[87] Tillich was probably referring to Friedrich Schleiermacher, who wrote his first major publication, *On Religion: Addresses to Its Cultured Despisers*, in an attempt to defend religion and to make it appealing to nonbelievers.[88] Schleiermacher said that religion was not based so much on revelation or dogmas as on a "feeling for the unity underlying all finite and temporal things." It is this feeling, he declared, that generates the idea of God.

Some similarity between Schleiermacher's and Einstein's philosophy of religion is to be expected because, like Einstein, Schleiermacher was greatly influenced by Spinoza. Schleiermacher called him "the holy excommunicated Spinoza . . . for whom the infinite was his beginning and his end, the universe his only and lasting love."[89] Schleiermacher's declaration that "our pious self-consciousness (frommes Selbstbewusstsein), by which we relate everything that stimulates or affects us to God, is identical with

[87] P. Tillich, "The Idea of a Personal God."

[88] F. Schleiermacher, *Über die Religion—Reden an die Gebildeten unter ihren Verächtern* (Brockhaus, Leipzig, 1868; first published in 1799).

[89] "Opfert mit mir ehrerbietig eine Locke den Manen des heiligen verstossenen Spinoza! Ihn durchdrang der hohe Weltgeist, das Unendliche war sein Anfang und Ende, das Universum seine einzige und ewige Liebe." Ibid., p. xi. See also T. Camerer, *Spinoza und Schleiermacher* (Cotta, Stuttgart, 1903), and S. Simon, *Spinozas Gottesidee in ihrer Wirkung auf das religiöse Empfinden Schleiermachers und Goethes* (Selbstverlag, Inowraclaw, 1937).

the recognition (Einsicht) that all this depends on, and is determined by, the unity of nature" is, in fact, quite similar to the basic idea of Einstein's cosmic religion.

It is not difficult to find other theologians who expressed similar ideas. A noteworthy example is Einstein's contemporary Albert Schweitzer, who in addition to his well-known humanitarian work as a physician and his musical talent as an organist was also a scholarly author of theological treatises. In some of his writings, he suggested the notion of an impersonal God, such as when he referred to God as the "Cosmic Will whence all life emanates."[90]

It is very likely that Einstein had read Schleiermacher,[91] it is less likely that he had read Schweitzer.[92] Because Einstein, as we know, had been influenced predominantly by Spinoza, the influence of Schleiermacher or any other more recent theologian would not have been important.

Because Einstein's philosophy of religion does not comprise mystical or theosophical conceptions, it is a rationalistic system of thought and therefore—almost like a scientific theory—can be subjected to methodological criteria of internal logical consistency as well as to Einstein's epistemological presuppositions. Although some separate theses of this system—such as its denial of a personal God—have been widely criticized, Einstein's philosophy of religion in its entirety has been given little attention, just

[90] A. Schweitzer, *Das Christentum und die Weltreligionen* (Beck, Munich, 1922, 1925; Haupt, Bern, 1948); *Christianity and the Religions of the World* (Allen and Unwin, London, 1939).

[91] A copy of Schleiermacher's *Über die Bildung zur Religion* (Diederich, Jena, 1916), which is an excerpt of *Über die Religion*, was in Einstein's library in Berlin, now at the Einstein Archive in Jerusalem.

[92] Einstein and Schweitzer corresponded with each other, but their correspondence dealt almost exclusively with social issues.

as his personal attitude toward religion has not been adequately considered by his biographers. Moreover, the few existing methodological analyses of his cosmic religion differ considerably in their conclusions.

Let us compare, for example, two critical articles on this issue that were published in the same journal and in the same year. Roy D. Morrison, professor of philosophical theology at Wesley Theological Seminary in Washington, D.C., analyzed Einstein's philosophy of religion within the context of the general conceptual strife between religion and science and argued that "for Einstein, there is a single attitude which lies at the base of religion, philosophy, and science. This attitude, which is religious in the highest sense of the word, motivates the striving for the highest ethical ideals and the striving for the deepest possible grasp of the intelligibility of the cosmos."[93] Morrison pointed out that, in contrast to Kant's transcendental deduction of a priori synthetic judgments and his differentiation between theoretical and practical reason, Einstein does not posit rigidity or proof for the basic postulates, but regards them as free inventions of the scientific imagination, justified by their operative success in providing intelligibility and supported by an attitude that is "akin" to religious faith. There exists therefore a complete unity between Einstein's scientific methodology and his nondogmatic conception of religion. Morrison concluded his article with the statement, "I believe that critical philosophy of religion now has the task of reassessing our religious heri-

[93] R. D. Morrison II, "Albert Einstein: The Methodological Unity Underlying Science and Religion," *Zygon—Journal of Religion and Science* 14 (1979): 255–266.

tage in the light of Einstein's work and its methodological unity."

Starting from the same premises, Dean R. Fowler, professor of theology at Marquette University, Milwaukee, Wisconsin, arrived at an almost diametrically opposite conclusion. He agreed with Morrison that, according to Einstein, scientific concepts are freely created inventions of our mind and that no logical connection exists between our sense experience and our theoretical concepts of the universe. But, Fowler pointed out, according to Einstein, religious feeling is based on the discovery of the rational structure of the universe. How then, asked Fowler, "do we know that our theories are in fact about the world?" Einstein's cosmic religion, Fowler continued, is also afflicted with a basic ambiguity. If the recognition of the rationality or intelligibility of the world leads Einstein to the conviction that there exists, as he declared, a "superior mind that reveals itself in the world of experience" and "this firm belief . . . represents my conception of God," then Einstein is a theist. If, however, "God" refers to the inherent structure of the universe itself, rather than to the source of this structure, then Einstein is a pantheist. Furthermore, the classical radical separation between mind and nature or subject and object, which lies at the basis of Einstein's cosmic religion, contradicts, according to Fowler, the implications of Einstein's scientific thought according to which the world is a field of interdependent events. "The world," Fowler concluded,

> is a complex web of interrelations wherein the categories of subject and object merge, blurring the traditional dualistic distinctions. Cosmic religion, however,

does not emerge from the epistemological and metaphysical insights of the contemporary scientific vision, a vision shaped largely by the Einstein revolution. Rather, cosmic religion is an attempt to understand the relationship between science and religion in the framework of outdated categories. It is an attempt, I believe, which is unsuccessful.[94]

Einstein's ideas about the relationship between science and religion also have been criticized, though on different grounds, by philosopher and theologian Frederick Pond Ferré, formerly at Dickinson College in Pennsylvania and now a research professor at the University of Georgia. Ferré contended that Einstein "impatiently" rejected the popular view that science has nothing to do with religion by maintaining that "an important nonreciprocal relationship holds between religion and science: science is greatly dependent upon religion, but not vice versa."[95] Ferré distinguished among six different ways in which, according to Einstein, religion promotes the scientific enterprise: the heuristic, epistemic, emotional, motivational, supportive, and personal. In order to substantiate this sixfold differentiation, Ferré documented each category with relevant citations from Einstein's writings. To vindicate his claim concerning the heuristic contribution of religion to science, Ferré referred to Einstein's denial of the existence of a logically coercive path from sense experience to the formulation of fundamental laws in science. Ferré quoted from

[94] D. R. Fowler, "Einstein's Cosmic Religion," *Zygon—Journal of Religion and Science* 14 (1979): 267–278.

[95] F. Ferré, "Einstein on Religion and Science," *American Journal of Theology and Philosophy* 1 (1980): 20–28.

Einstein's preface to a book by Max Planck, "the supreme task of the physicist is the discovery of the most general elementary laws from which the world picture can be deduced logically. But there is no logical way to the discovery of these elemental laws."[96] To show that religion does the job of spanning the gap, Ferré cited Einstein's 1930 interview with James Murphy and J. W. N. Sullivan: "I [Einstein] am of the opinion that all finer speculations in the realm of science spring from a deep religious feeling." This, then, is Ferré's explication of Einstein's conception of the heuristic contribution of religion to science.

Instead of discussing Ferré's analogous explications of the other five ways, we concentrate on his contention that, according to Einstein, "the influence of religion, functionally understood, upon science is not reciprocated by science upon religion." In order to prove this contention, Ferré quoted from Einstein, "While it is true that science, to the extent of its grasp of causative connections, may reach important conclusions as to the compatibility and incompatibility of goals and evaluations, the independent and fundamental definitions regarding goals and values remain beyond science's reach."[97] What seems to have influenced Ferré most was probably Einstein's answer to Murphy, "I think you are right in speaking of the moral foundations of science; but you cannot turn it around and speak of the scientific foundations of morality."[98]

Is Ferré's nonreciprocity contention really defensible? Has

[96] A. Einstein, Preface to M. Planck, *Where is Science Going?* (W. W. Norton, New York, 1932), p. 12.

[97] A. Einstein, "Religion and Science: Irreconcilable?"; see also "The Goal."

[98] A. Einstein, *Science and God*, p. 375.

not Einstein explicitly declared: "[E]ven though the realms of religion and science in themselves are clearly marked off from each other, nevertheless there exist between the two strong reciprocal relationships and dependencies"?[99] And does not his often quoted metaphor "Science without religion is lame, religion without science is blind"[100] indubitably affirm his belief in a reciprocity of the relation between science and religion? Moreover, because blindness is certainly a more severe handicap than lameness, one can even say, in contrast to Ferré, that Einstein may have regarded the impact of science on religion of much greater importance than the impact of religion on science.

Did Einstein contradict himself by affirming reciprocity in the statements quoted above while denying it in those quoted by Ferré? He did not! Ferré's quotations from Einstein deal exclusively with "goals and values" or "moral philosophy," notions that have little to do with Einstein's conception of cosmic religion. According to Einstein, "there is nothing divine about morality,"[101] or as he once wrote to a community leader in the Boston area, "morality has nothing to do with religion."[102] In fact, Ferré's concepts of religion, which he elsewhere defined as "one's way of *valuing* most comprehensively and intensively," differs fundamentally from Einstein's conception of this term.[103]

[99] A. Einstein, "Science and Religion"; *Out of My Later Years*, p. 29.

[100] Ibid., p. 30.

[101] A. Einstein, *Ideas and Opinions*, p. 40; *The World as I See It*, p. 28; see also Einstein to Solovine, 1 January 1951.

[102] Einstein to R. E. Benenson, 31 January 1946; Einstein Archive, reel 33-317.

[103] F. Ferré, *Basic Modern Philosophy of Religion* (George Allen and Unwin, London, 1967), p. 69 [emphasis added].

In the sequel, Ferré discussed how the belief in strict determinism affected Einstein's philosophy of religion no less than his physical thought, Although we deeply appreciate Einstein's intense religiousness, Ferré continued,

> there remain important unanswered questions. For one thing, where would Einstein . . . draw the boundaries between faith and reason in matters of scientific theorizing? Granting that religious insights may be essential to the scientific good, how long may an individual scientist hold out against a scientific concensus, on religious grounds, without becoming unreasonable? [Here Ferré seemed to have in mind Einstein's persistent opposition to the standard interpretation of quantum mechanics.] . . . How can a position like Einstein's deal with the sense possessed by many religious geniuses of the enormous value of individual human life or the ultimate poignancy of historical choice? Must these value intuitions be merely denied or set aside as inferior? Einstein's cosmic religion may, ironically, not be large enough to weave together all the genuine values of religious and moral intuitions. Adequacy to the full data of experience is no less needed in the domain, of religion, where values predominate, then in that of science.

Comparing Einstein's philosophy of religion with his philosophy of science, Ferré charged Einstein with having disregarded, to some extent at least, not only the full data of experience but also his own methodological criteria for a good theory as he had formulated them in his "Autobiographical Notes": "A theory is the more impressive the greater the simplicity of its premises is, the more different kinds of things it relates, and the more extended is its area of applicability." Had Einstein applied these criteria to his religious theory as he had applied them to his physical theory he would have recognized, Ferré declared, that his cosmic religion had to be broadened to include also an appreciation of moral values and of personal freedom.

In a letter, of June 18, 1998, addressed to the author, Professor Ferré summarized his critique of Einstein's religious philosophy as follows:

> I agree that Einstein did not consider morality divine. Unfortunately, on his cosmic religion, taken at full strength, it is hard to see any respectable place at all for personal morality. And this is odd, coming from a man so obviously moved by great moral impulses concerning war, Israel, etc. This is another example, I think, of the need to broaden cosmic religion still further to include an honored place for the personal and freedom, without which morality is illusion.

UNLIKE THE essays just discussed, the following summary does not review Einstein's religious philosophy by treating it as a complete system of thought, but rather reconstructs it by tracing step-by-step the intellectual development that led to his cosmic religion. Although special attention is paid to the intellectual incentives and psychological motives that were active in the early stages of this process, this section is not merely a summary of the preceding discussions. Because part of its purpose is to assign Einstein's religion to its proper place within the spectrum of religious beliefs and to amend an unfortunately common misinterpretation of Einstein's philosophy of religion, this summary includes some important comments that have not yet been brought to the reader's attention.

A truthful reconstruction of the different phases of this development must be based, of course, on documentary evidence, but Einstein has never written what may be called a *religious* autobiography. His "Autobiographical Notes" also are not very helpful. They briefly discuss only a single episode in this process, namely how, as a result of reading certain popular scientific books, Einstein abandoned his juvenile religiosity and became a freethinker until certain philosophical considerations changed his mind.[104] Finally, to extract a coherent account of the development of his religious philosophy from his numerous writings, would be a difficult and tedious enterprise.

A more convenient and, as we shall see, quite reliable method consists in referring to Einstein's correspondence with those who asked him to define his religious credo. For Einstein has so carefully worded and so logically formulated his answers that they can guide us in this task. A typical example is his 1947 letter to Murray W. Gross of Brooklyn, New York. Gross had written to Einstein on behalf of his grandfather, an aged Talmudic scholar and author of a rabbinical treatise, who was eager to know Einstein's views about God and religion. Einstein replied:

> It seems to me that the idea of a personal God is an anthropomorphic concept which I cannot take seriously. I feel also not able to imagine some will or goal outside the human sphere. My views are near to those of Spinoza: admiration for the beauty of and belief in the logical simplicity of the order and harmony which

[104] A. Einstein, "Autobiographical Notes," p. 5.

> we can grasp humbly and only imperfectly. I believe that we have to content ourselves with our imperfect knowledge and understanding and treat values and moral obligations as a purely human problem—the most important of all human problems.[105]

It is not adventitious that the letter, like many other letters of this kind, begins with the rejection of the use of an anthropomorphism in religious thought. For this rejection, I contend, marked also the beginning of Einstein's religious thinking and continued to play a dominant role at its later stages as well. In order to prove this contention, I must elaborate on this issue at some length.

Recall Moszkowski's statement that when Einstein was at school, he "clung to a faith free of all doubt . . . and read the Bible without feeling the need of examining it critically."[106] What could have been the object of such a critical examination? In his Bible lessons, young Albert certainly read verses like these: "God is walking in the garden" (Genesis 3:8), and "smelled the pleasing odor" of sacrifices (Genesis 8:21), "God spoke to Moses" (repeatedly in Leviticus), the tables of testimony "were written with the finger of God" (Exodus 31:18; Deuteronomy 9:10), or God "was sorry that He had made man on earth" (Genesis 6:6). These are, of course, only a few of the anthropomorphisms and anthropopathisms with which the Bible, like many sacred books of other religions, is replete. On the other hand, young Albert certainly also read the second commandment

[105] Einstein to M. W. Gross, 26 April 1947. Einstein Archive, reel 58-243. Einstein's letter to M. Magalaner, also of 26 April, 1947 (reel 33-324), is almost identical with this letter to Gross.

[106] A. Moszkowski, *Einstein the Searcher*, p. 221.

of the Decalogue, "You shall not make yourself an image or any likeness of God" (Exodus 20:4; Deuteronomy 5:8), and, in the prayer book, the third of Maimonides' thirteen Articles of Faith, "I firmly believe that the Creator, blessed be His name, is not corporeal; that no bodily accidents apply to Him; and that there exists nothing whatever resembles Him." According to Moszkowski, the young Einstein, in his religious fervor, did not yet recognize that literal interpretations of such statements raised a serious theological problem.

As reported by Einstein himself in his "Notes," he "became a fanatic freethinker" as soon as he began reading popular scientific books. He declared that this attitude "has never again left me, even though later on, because of a better insight into the causal connections, it lost some of its original poignancy."

We know that these "popular scientific books" were those that Max Talmey had advised him to read: Aaron Bernstein's *Popular Books on Physical Science* and Ludwig Büchner's *Force and Matter*. A perusal of these books shows that, with the exception of three chapters that severely criticize anthropomorphisms in religious thought, they deal almost exclusively with purely scientific topics unrelated to any religious issues.[107] Büchner's critique begins with a historical survey of theriomorphic and therianthropic religions in which animals or, respectively, combi-

[107] In A. Bernstein's *Naturwissenschaftliche Volksbücher* (F. Dümmler, Berlin, 4th ed. n.d.), these are chapter 12 (pp. 49–53) and chapter 13 (pp. 53–57) of volume 12; in L. Büchner's *Kraft und Stoff* (T. Thomas, Leipzig, 9th ed., 1867), it is the chapter entitled "Die Gottes-Idee" (pp. 191–201).

nations of animals and humans are apotheosized, and it concludes with Xenophanes' well-known statement, "If oxen or lions had hands and could paint, they would undoubtedly represent their gods in the form of oxen or lions." It is psychologically understandable that the reading of such texts, at the impressionable age of 12, had a profound impact on Einstein's mind. It led him to the conclusion that "much in the stories of the Bible could not be true." He consequently lost faith in the Bible and renounced his erstwhile religiosity as naive. We can therefore safely conclude that Einstein's more mature thinking about religion began with an intense aversion to anthropomorphisms. This aversion continued to dominate his mind as shown by his letter to Gross, written at age 68, that begins with the rejection of an anthropomorphism.

Einstein's notion of "anthropo*morphism*" as he applied it to "the idea of a personal God" included also the notion of "anthropo*pathism*," a term he never used. As the second sentence of the letter shows, he rejected the ascription to God not only of human physical traits or "form" (*morphē*) but also of human feelings or intentions (*pathos*), as his denial of "will and goal outside the human sphere" indicates.

Einstein had many precursors—besides Bernstein and Büchner—in opposing anthropomorphism in religion. The most famous was undoubtedly Moses Maimonides, the foremost Jewish philosopher of the Middle Ages and the classical exponent of rationalism in religious philosophy. Even today he is still highly respected as a leading rabbinical authority. He condemned anthropomorphism not only in the above-quoted Article of Faith but with even greater fervor in his monumental *Guide for the Perplexed*, published in 1190 in Arabic under the title *Dalalat al Hai'rin.*[108] According to Maimonides, the three fundamental truths of religion are the existence, unity, and incorporeality of God; the affirmation of any other positive attribute would contradict His unity. Hence, any positive statement about God must be interpreted as the denial of the corresponding negative statement.[109] Only an indirect knowledge of God is therefore possible. In order to explain how this imperfect knowledge of God is accessible, Maimonides quoted Moses's request to God (Exodus 33:13), "Show me now thy ways and I shall know thee," and interpreted it as indicating that God can be known only indirectly by knowing His "ways," that is, His actions or manifestations in Nature.[110] Referring to these statements, a modern expert on medieval philosophy declared: "Evidently the idea of God [as conceived by Maimonides] loses in this manner much of its personal character."[111] In fact, Maimonides' conception of God and Einstein's denial of a personal God, although certainly not identical, have much in common, and the public reactions to them were also quite similar. The uproar caused by Einstein's 1940 paper "Science and Religion," in which he argued for the impersonality of God, had a precedent in the

[108] Of course Maimonides did not yet use the term "anthropomorphism," which was not coined until the middle of the eighteenth century. He used Arabic terms equivalent to "equivocality," "amphiboly," and "figurativeness."

[109] H. A. Wolfson, "Maimonides on Negative Attributes," *Jubilee Volume for Louis Ginzberg* (American Academy for Jewish Research, New York, 1945), pp. 411–446.

[110] Maimonides, *The Guide of the Perplexed,* book 1, chap. 54.

[111] J. Guttmann, "Commentary," in Maimonides, *The Guide of the Perplexed,* abridged ed. Trans. C. Rabin (East and West Library, London, 1952), p. 210.

protestations against Maimonides's teachings by orthodox fundamentalists, especially in southern France.[112]

It would be incorrect, however, to assume that Einstein had been influenced by Maimonides. True, at the 1935 Maimonides Jubilee Celebration, Einstein said that Maimonides had exerted "a crucial and fruitful influence" on both his contemporaries and also "on later generations." But Einstein could not count himself among those whose work had been influenced by Maimonides. He never referred to Maimonides in his writings; and he read him probably only in the last years of his life.[113]

Another similarity of Einstein's philosophy of religion with that of Maimonides is worth mentioning. If the process of purging religious thought of anthropomorphisms is carried out to its logical extreme by not allowing God any attribute whatever—and not only a human one, for every attribute we can think of is a human concept—then only two alternatives are left: atheism, the denial of the existence of God, no matter how "God" is defined; or mysticism, a "mystical silence" in an unmediated communion of man's soul with the highest reality.[114] Maimonides, the ra-

[112] E.g., in 1233 Rabbi Solomon of Montpellier succeeded to persuade church authorities to anathematize the *Guide* as a heretical work and to consign it publicly to the flames. The tendency in the *Guide* to harmonize Aristotelian philosophy with theistic religion played an additional role in this condemnation.

[113] "Leider habe ich Maimonides nicht gelesen" ("Unfortunately I have not read Maimonides"), Einstein to A. Geller, 4 September 1930, Einstein Archive, reel 46-362. The only work by, or on, Maimonides that was in Einstein's possession was a 1946 publication of M. Friedländer's 1881 English translation of the *Guide*, now kept in the Collection of Einstein's books at the Einstein Archive in Jerusalem.

[114] See, e.g., R. J. Z. Werblowski, "Anthropomorphism," in M. Eliade,

tionalist, avoided both alternatives by arguing on the one hand that "the highest knowledge of God consists of knowing that we are unable to comprehend Him" because no positive attributes can be ascribed to Him, and on the other hand, that we know His existence because it can be inferred from His "ways" or actions as manifested in Nature. In a similar vein, Einstein declared that there is no personal God, but "every one who is seriously involved in the pursuit of science becomes convinced that a spirit is manifest in the laws of the Universe—a spirit vastly superior to that of man, and one in the face of which we with our modest powers must feel humble."[115]

Coming now to the third sentence of the letter to Gross where Einstein declared that his "views are near to those of Spinoza," we know that, in contrast to the *Guide*, Spinoza's *Ethics* had been read repeatedly by Einstein. He had read it in the early years in Bern and certainly again, together with Spinoza's *Correspondence*, in the late twenties, as we know from a letter he wrote in 1928 to Leo Szilard. Because Spinoza was his favorite philosopher, Einstein had probably read his works many times over. But having read an author or even admitting a similarity of views with him does not necessarily mean to have been influenced by him. What role then, precisely, did Spinoza play in the development of Einstein's religious philosophy? In order to answer this question, it is first necessary to comment on the *Ethics* itself.

In a discussion at the Sorbonne with French philoso-

---

ed., *The Encyclopedia of Religion* (Macmillan, New York, 1987), pp. 318–319.

[115] H. Dukas and B. Hoffmann, *Albert Einstein—The Human Side*, p. 33.

phers in 1922, Einstein said that "every philosopher has his own Kant." He could have said the same of Spinoza's *Ethics*. In order to see in how many different ways the *Ethics* has been interpreted, it suffices to recall the following facts. It was published posthumously, but not when it was completed in 1675, because of the intervention of certain theologians who regarded it as a heresy. David Hume still spoke of it as the work of an "atheist." Later, men of letters such as Lessing, Goethe, Herder, Coleridge and Shelley, praised it as unsurpassed in sublimity and profundity; Schleiermacher asked that we "offer a lock of hair to the manes of the holy and excommunicated Spinoza."[116] His contemporary, the romantic poet Novalis even called the *Ethics* the work of a "god-intoxicated man."[117] How could Spinoza's magnum opus become the object of such widely diverging evaluations?

As its full title, *Ethica ordine geometrico demonstrata*, indicates, the *Ethics* is a philosophical treatise formulated à la Euclid's *Elements* as an axiomatic system: It is based on definitions and axioms from which propositions are derived by purely logical deductions. In every axiomatization, the basic terms, even if formally defined, allow a certain latitude in interpretation. Euclid defined a point as "that which has no part," but he never used this definition when proving a theorem. One can therefore say, using the terminology of modern logicians, that the primitive terms of axiomatised geometry are implicitly defined by the

[116] F. Schleiermacher, *Über die Religion*, p. XI. In note 88 the German original of Schleiermacher's work is quoted. Cf. note 89.

[117] "Ein gottrunkener Mensch." Novalis (Freiherr Friedrich von Hardenberg), *Gesammelte Werke* (Bühler, Herrliberg-Zürich, 1946), vol. 4, p. 259.

function they play in the axioms or propositions. Logicians call an axiomatic system "categorical" if all its interpretations (or "models") are isomorphic. In Spinoza's axiomatized *Ethics*, the situation is different. The important term "substance" (substantia), for example, is defined as that "which is in itself and is conceived through itself" [quod in se est et per se concipitur]; and the proof of the first proposition, "the substance is by nature prior to its affections" [substantia prior est natura suis affectionibus], makes explicit and critical use of the definition of substance. Hence, if the definitions lack categoricity, the whole axiomatic system lacks categoricity. For a scholar who, like H. A. Wolfson, has profoundly studied the historical development of Spinoza's philosophical terminology from Aristotle to Descartes, the definitions in the *Ethics* may have a high degree of categoricity.[118] For other readers, they may not. It is, of course, this lack of categoricity that is the reason for the above-mentioned wide divergence of judgments.

Einstein seems to have been aware of this fact when he declared in his *Foreword* to a *Spinoza Dictionary* that "everyone may interpret Spinoza's text in his own way,"[119] or when he told Hessing that "to love Spinoza does not suffice to be allowed to write about him: this one must leave to those who have gone further into the historical background."[120]

That Einstein loved and admired Spinoza cannot be doubted; suffice it only to recall the poem he wrote in

[118] H. Wolfson, *The Philosophy of Spinoza*, vol. 1.

[119] A. Einstein, Foreword to D. D. Runes, *Spinoza Dictionary* (Philosophical Library, New York, 1951), p. vi.

[120] Einstein to S. Hessing, 8 September 1932.

honor of Spinoza. It is equally well known that he saw in him a kindred spirit and that his "views are near to those of Spinoza," as seen in the letter to Gross. To feel close to Spinoza and to sympathize with him does not necessarily mean to have been influenced by him. How far, then, was Einstein's intellectual development influenced by Spinoza, if at all?

This is not a trivial question. Einstein's "Autobiographical Notes," which are a review of his intellectual life, do not contain a single reference to Spinoza or to the *Ethics*. In fact, Spinoza's name is not mentioned, whereas Hume and Kant, for example, are referred to repeatedly. In his other writings, Einstein has never quoted a specific statement by Spinoza as a source or inspiration for an idea of his own. In general, Einstein found in Spinoza—and only in Spinoza—a courageous and eloquent propounder of the same ideas he had thought to be correct before, or independently of, his reading the *Ethics*. For example, when Einstein was in his early twenties in Bern and read the *Ethics* for the first time, he must have been impressed by Spinoza's treatment of philosophical problems *more geometrico*, as well as delighted to find support for his own beliefs in the Scholium to proposition 15, where Spinoza called anthropomorphisms "the greatest absurdity" [nihil absurdius de Deo . . . dici potest]. It is legitimate to say that Spinoza influenced Einstein, but this influence expressed itself mostly in strengthening and articulating conceptions that had previously been germinating in Einstein's mind. This also seems to apply to the idea of a rigid determinism or causality in the physical universe, an idea of cardinal importance in the philosophies of Spinoza and of Einstein. However, according to Einstein, Spinoza was the first to conceive of extending this idea to an all-pervasive determinism to human thought, feeling, and action.[121] This extension was a concept that Einstein inherited from Spinoza without having anticipated it. As explained above, the motivation for introducing the cosmological constant may well have been the direct result of Spinoza's influence.[122] I do not agree with the suggestions that the *Ethics* influenced Einstein's physical thinking in general as, for example, Boris Kouznetsov's claim that Einstein's notion of the unified field "is the analogue of Spinoza's substance"[123] or, as Sylvaia Zac sweepingly stated, "Spinoza's metaphysics heralds Einstein's physics."[124]

To sum up, Einstein, like Maimonides and Spinoza, categorically rejected any anthropomorphism in religious thought. Like Spinoza, Einstein regarded the idea of a personal God as an anthropomorphism. Unlike Spinoza, who saw the only logical consequence of the denial of a personal God in an identification of God with Nature [deus sive natura], Einstein maintained that God manifests himself "in the laws of the Universe as a spirit vastly superior to that of man, and one in the face of which we with our modest powers must feel humble." Einstein agreed with Spinoza that he who knows Nature knows God, but not because Nature is God but because the pursuit of science

[121] Einstein to D. Runes, 8 September 1932.

[122] Spinoza, *Ethics*, col. 2 to proposition 20, pt. 1.

[123] "Il est evident que le champ unifié . . . est l'analogue de la substance de Spinoza." B. Kouznetsov, "Spinoza et Einstein," *Revue de Synthèse* 88 (1967): 38.

[124] S. Zac, "On the idea of creation in Spinoza's philosophy," in Y. Yovel, *God and Nature: Spinoza's Metaphysics* (E. J. Brill, Leiden, 1991), p. 240.

in studying Nature leads to religion. In the terminology of theology, Einstein's religion may therefore be called a naturalistic theology according to which knowledge of God can be obtained by observing the visible processes of nature, but with the proviso that the manifestation of the divine in the universe is only partially comprehensible to the human intellect.

Einstein's cosmic religion is, of course, incompatible with the doctrines of Judaism, Christianity, Islam, and all other theistic religions. The crucial difference lies in its denial of a personal God who punishes the wicked or rewards the righteous and performs miracles by breaking the causal laws of nature. Petitionary prayers have therefore no place in Einstein's religion. "As long as you pray to God and ask him for some benefit," he once wrote to Szilard, "you are not a religious man." It is therefore not surprising that Einstein's denial of a personal God encountered strong opposition, especially on the part of the clergy as, for example, at the 1940 New York Conference.[125] It is also not surprising that Einstein was called an atheist. But Einstein never regarded his rejection of a personal God as a denial of God. Recall his reaction to Büsching's book entitled *Es gibt keinen Gott* (*There Is No God*) in which he declared that a belief in a personal God seems "preferable to the lack of any transcendental outlook of life."[126] In spite of his denial of a personal God and his rejection of religious customs and rituals, he had a high respect for traditional religion. "The highest principles for our aspirations and judg-

[125] See notes 39 to 45. For example, E. F. W. to Einstein, 12 September 1940 (n. 48).

[126] Einstein to E. Büsching, 25 October 1929.

ments," he declared in 1939, "are given to us in the Jewish-Christian religious tradition. It is a very high goal which, with our weak powers, we can reach only very inadequately, but which gives a sure foundation to our aspirations and valuations."[127] He also venerated the founders of the great religions, as can be seen from a message he sent to the National Conference of Christians and Jews in 1947: "If the believers of the present-day religions would earnestly try to think and act in the spirit of the founders of these religions, then no hostility on the basis of religion would exist among the followers of the different faiths. Even conflicts in the realm of religion would be exposed as insignificant."[128]

Of course, an atheist can also respectfully acknowledge the values and merits of theistic religions. After having gained what he called "a better insight into the causal connections," Einstein always protested against being regarded as an atheist. In a conversation with Prince Hubertus of Lowenstein, for example, he declared, "What really makes me angry is that they ["people who say there is no God"] quote me for support of their views."[129] Einstein renounced atheism because he never considered his denial of a personal God as a denial of God. This subtle, but decisive, distinction has long been ignored. The lack of regard for this distinction some years ago led the *Osservatore Romano*, the organ of the Roman Catholic Church, to proclaim that Einstein's cosmic religion is "an authentic atheism even if it is camouflaged as cosmic pantheism," a

[127] A. Einstein, "The Goal," p. 43.

[128] H. Dukas and B. Hoffmann, *Albert Einstein—The Human Side*, p. 96.

[129] R. W. Clark, *Einstein—The Life and Times*, p. 516.

statement that was later retracted.[130] A more recent example is the curt statement that "he [Einstein] was a lifelong atheist" from the Foreword to a popular book on Einstein's life, published in 1998 by the Einstein Archive in Jerusalem. That statement is obviously contradicted by another statement by Einstein quoted in the same book, "The divine reveals itself in the physical world."[131] In fact, when Einstein declared that "science without religion is lame, religion without science is blind," a statement that summarizes his religious credo, he did not use the term "religion" to mean "atheism."[132]

[130] Quoted in S. S. Kantha, *An Einstein Dictionary* (Greenwood Press, Westport, Conn., 1996), p. 146.

[131] Z. Rosenkranz, *Albert through the Looking-Glass* (Jewish National and University Library, Jerusalem, 1998), pp. xi, 80.

[132] Not only was Einstein not an atheist, but his writings have influenced people to turn away from atheism, although he undoubtedly had never intended to convert anybody to his own conviction. He discussed religion only in response to requests by people who asked him about his religious outlook, as in the case of the letter to Gross, or when he was asked to explain his view in journals or in religious conferences. Nevertheless, after publication of the brief German edition of the present book, the author received some letters—mostly from scientists and including an internationally well-known biophysicist—in which the writers admitted that they had been atheists until they read Einstein's concept of religion, which inspired them to become deeply religious. Needless to say, the author was greatly surprised by these letters because the book, as strongly emphasized in the Introduction, had been designed to serve merely as a historical account of Einstein's religious thinking without any intention of converting its readers to Einstein's view or interfering in any way with their religious beliefs.

CHAPTER 3

Einstein's Physics and Theology

The statement "science without religion is lame, religion without science is blind," with which Einstein epitomized his philosophy of religion, is stronger than even Ralph Waldo Emerson's trenchant aphorism "the religion that fears science, insults God and commits suicide."[1] According to Einstein's maxim, religion is not only compatible with science, it is also promoted by science just as it promotes science by stimulating and sustaining scientific research as exemplified by Kepler and Newton. The fundamental tenet of Einstein's cosmic religion is that science furthers religion. According to Einstein, religion is nurtured by the feeling of awe and reverence that accompany the discovery of the laws of nature and the awareness of the harmony that rules the universe.

The theory that studies par excellence the structure of the universe was for Einstein undoubtedly his theory of relativity with its implications for relativistic cosmology. Therefore, it may seem strange that when Randall Thomas Davidson, the Archbishop of Canterbury and head of the Anglican Church, asked "what effect relativity would have on religion," Einstein replied, "None. Relativity is a purely scientific matter and has nothing to do with religion."[2]

[1] Quoted in E. H. Cotton, ed., *Has Science Discovered God?* vi.

[2] Quoted after A. S. Eddington, *The Philosophy of Physical Science* (Cambridge University Press, Cambridge, 1939), p. 7; quoted after P. Frank, pp. 189–190. According to R. W. Clarke, the Archbishop's question referred to "morale" and not to "religion." But it seems that the version reported by Eddington and Frank is more authentic than Clarke's, because Eddington participated at the meeting between Davidson and Einstein, and Frank based his report on personal communications with Einstein.

Eddington did not agree with Einstein's reply. According to Eddington,

> those who quote and applaud the remark as though it were one of Einstein's most memorable utterances overlook a glaring fallacy in it. Natural selection is a purely scientific theory. If, in the early days of Darwinism, the then Archbishop had asked what effect the theory of natural selection would have on religion, ought the answer to have been "None. The Darwinian theory is a purely scientific theory, and has nothing to do with religion"? The compartments into which human thought is divided are not so watertight that fundamental progress in one is a matter of indifference to the rest. The great change in theoretical physics which began in the early years of the present century is a purely scientific development; but it must affect the general current of human thought, as at earlier times the Copernican and Newtonian systems have done.[3]

The conversation with Davidson took place about ten years before Einstein wrote his essays on cosmic religion, which are in full agreement with Eddington's conception of the relation between science and religion. It may therefore be possible that Einstein had changed his mind in the intervening years. It seems more likely that Einstein had not committed that "glaring fallacy." His fallacy was probably only an equivocal use of the term "religion." He used the term "religion" in the sense of "religiosity" in his essays and in the sense of "institutional religion" or "orga-

[3] Quoted in E. H. Cotton, ed., *Has Science Discovered God?*

nized religion" in his reply to the Archbishop. This interpretation is supported because, when asked to define his religion, Einstein apparently used a paradox by calling himself a "deeply religious nonbeliever," where "religious" obviously referred to "religiosity" and "nonbeliever" referred to his refusal to belong to an "institutional religion" or "religious community."[4]

However, the subject matter of this chapter is not whether science promotes religiosity or whether science can induce a person to join a religious denomination. Rather, the question is whether Einstein's "purely scientific" work has any implications that are relevant to religious issues or to theology as a systematic discipline.

THE IDEA of drawing theological consequences from physics has a long history. Classical physics had long been invoked to defend, perhaps more than to attack, religious doctrines. Isaac Newton, for example, repeatedly stated that it was part of the task of natural philosophy, as physics was called in his time, to discuss questions concerning the attributes of God and his relationship to the physical world. Having thought to have empirically established—for instance, by means of his famous pail experiment—the existence or reality of absolute space, he did not hesitate to identify it, in the *Scholium Generale* of his *Philosophiae Naturalis Principia Mathematica,* with God's omnipresence, and absolute time with God's eternity. In *Query 28* of his *Opticks,* moreover, he concluded from physical "phenomena

[4] "Man wird zum tief religiösen Ungläubigen. (Dies its eine einigermassen neue Art von Religion)." Einstein to H. Mühsam, 30 March 1954. Einstein Archive, reel 38-434.

that there is a Being incorporeal, living, intelligent, omnipresent, who in infinite space, as it were in His Sensory, sees the things themselves intimately, and thoroughly perceives them, and comprehends them wholly by their immediate presence to Himself." Newton's letters to Richard Bentley likewise abound with arguments that the structure of the universe implies the existence of God as Creator.[5]

With the appearance of William Paley's book *Natural Theology* in 1802, the idea that the study of nature and of physics in particular reveals a world that conceals a divine intelligence gained wide recognition and continues to do so today.[6] In a recent example, the well-known physicist and Nobel laureate Charles Townes declared in 1995, "Physicists are running into stone walls of things that seem to reflect intelligence at work in natural law."[7] More closely related to our topic and yet reminiscent of Newton's words is the following passage from *The Encyclopedia of Religion:* "Undoubtedly, Einstein's general relativity and the scientific cosmology it inspired are a classical case of creative science at its best. That it strongly supports metaphysics at

[5] See, e.g., the collection of Newton's letters to Bentley in H. S. Thayer, *Newton's Philosophy of Nature* (Hafner, New York, 1953), pp. 46–58. For details on the relation between theories of space and religion in the writings not only of Newton but also of his predecessors—the Italian natural philosophers like Francesco Patrizi or Tommaso Campanella, or the Cambridge Platonists like Henry More or Robert Fludd—see M. Jammer, *Concepts of Space: The History of Theories of Space in Physics* (Harvard University Press, Cambridge, Mass., 1954, 1969; enlarged, rev. ed., Dover, New York, 1993), chap. 3, 4.

[6] W. Paley, *Natural Theology: or Evidence of the Existence and Attributes of the Deity, Collected from the Appearances of Nature* (Faulder, London, 1802).

[7] C. Townes, *Making Waves* (American Institute of Physics, New York, 1995).

its best, which is the intellectual inference of the existence of a Creator, is but a replay of a now fairly old pattern of science."[8]

Einstein's theory of relativity led to a revision of the classical notions of space and time. Erwin Schrödinger, the discoverer of wave mechanics, intimated that relativistic notions of space and time also may have theological implications. In his 1956 Tarner Lecture, Schrödinger spoke about "the great stir both among the general public and among philosophers" that Einstein's special theory of relativity had aroused. He said,

> I suppose it is this, that it meant the dethronement of time as a rigid tyrant imposed on us from outside, a liberation from the unbreakable rule of "before and after." For indeed time is our most severe master by ostensibly restricting the existence of each of us to narrow limits—70 or 80 years, as the Pentateuch has it. To be allowed to play about with such a master's programme believed unassailable until then, to play about with it albeit in a small way, seems to be a great relief, it seems to encourage the thought that the whole "timetable" is probably not quite as serious as it appears at first sight. And this thought is a religious thought, nay I should call it *the* religious thought.[9]

When Schrödinger spoke about being liberated from the "unbreakable rule of 'before and after'" in classical physics, he was referring to Einstein's replacement of that rule by

[8] S. L. Jaki, "Science and Religion," in *The Encyclopedia of Religion*, vol. 13 (Macmillan, New York, 1987), p. 132.

[9] E. Schrödinger, *Mind and Matter* (Cambridge University Press, Cambridge, 1959), p. 82.

the relativity of the temporal sequence of two events that are causally not connectable, or in physical parlance, of two events whose space-time separation is spacelike. In other words, whether, under such conditions, event e_1 occurs before, simultaneously with, or after event e_2 depends on the choice of the reference frame in which they are observed. Indeed, the relativity of temporal order has been invoked to resolve certain theological problems as shown below. Another aspect of time, which has been used for the same purpose, is its relativistic conception as a coordinate in Minkowski's four-dimensional space-time, at least when the latter has been interpreted—as, for example, by Hermann Weyl—as a "block universe." As Weyl phrased it, "the objective world simply *is*, it does not *happen*. Only to the gaze of my consciousness, crawling upward along the lifeline of my body, does a section of this world come to life as a fleeting image in space which continuously changes in time."[10] In other words, the relations "earlier," "simultaneous with," and "later" are merely geometrical relations in the static four-dimensional space-time, and the terms "past," "present," and "future" have no objective reality.[11]

Whether the idea of a "block universe" is a logical consequence of the theory of relativity, or even only compatible with it, is not our present concern. It should be clear, however, that such a conception of the universe would seriously conflict with the Judeo-Christian religious tradition,

[10] H. Weyl, *Philosophy of Mathematics and Natural Science* (Princeton University Press, Princeton, N.J., 1949), p. 116.

[11] See also the differentiation between the "A-series" and the "B-series" in J. M. L. McTaggart, "The Unreality of Time," *Mind* 17 (1908): 457–474 and in his *The Nature of Existence*, vol. 2 (Cambridge University Press, Cambridge, 1927), chap. 33.

which assigns to time a very active role in history. It should be mentioned in this context that Einstein himself seems to have once embraced this idea and even attached to it a consolatory, almost religious connotation. When he was informed that his lifelong friend Michele Besso had died on March 15, 1955—four weeks before Einstein's death on April 18—he wrote to Besso's family, "Now he has departed a little ahead of me from this quaint world. This means nothing. For us faithful physicists, the separation between past, present, and future has only the meaning of an illusion, though a persistent one."[12]

As seen below, many more features of Einstein's special and general relativity have been invoked to resolve theological problems or to support or to refute certain religious theses. These theological implications are not listed in the chronological order in which they appeared in the literature; rather, they are discussed in more or less the same order in which the theorems to which they refer evolved in Einstein's step-by-step construction of his theory of relativity. Because textbooks on relativity introduce the theory in a way that differs considerably from Einstein's approach, a brief review of the genesis of the special theory of relativity follows.

THE THEORY of relativity owes its inception to Einstein's attempt to resolve the apparent contradiction between Gal-

[12] "Nun ist er mir auch mit dem Abschied von dieser sonderbaren Welt ein wenig vorausgegangen. Dies bedeutet nichts. Für uns gläubige Physiker hat die Scheidung zwischen Vergangenheit, Gegenwart und Zukunft nur die Bedeutung einer wenn auch hartnäckigen Illusion." Einstein to Vero and Mrs. Bice, 21 March 1955. Einstein Archive, reel 7-245; reprinted in *Albert Einstein—Michele Besso Correspondance 1903–1955* (Hermann, Paris, 1972), pp. 537–538.

ileo's so-called relativity principle, which states that all laws of physics are the same in all inertial reference frames, and the so-called light principle, required by Maxwell's electromagnetic theory and empirically verified by Michelson's famous experiment, which states that the velocity of light in a vacuum is the same in all inertial frames. These principles seemed to be incompatible because, if, in an inertial frame S_1, the velocity of light is c, then, in a frame S_2 moving relative to S_1 with constant velocity v, say, in the direction of the light beam, the velocity of light, according to the classical composition rule of velocities, should be $c - v$ and not c, as required by the light principle. Because the presupposition that causes this conflict is the classical law of adding velocities, and this law in turn is based on the doctrine of absolute time, Einstein realized that time must be relative; that is, different frames have different times. This requires that we must be "quite clear as to what we understand by 'time.' We have to bear in mind that all propositions involving time are always propositions about simultaneous events."[13] Therefore, the concept of simultaneity of spatially separated events started the whole development. In fact, the first paragraph of Einstein's 1905 seminal paper on relativity carries the heading "Definition of simultaneity."

[13] "Wir haben zu berücksichtigen, dass alle unsere Urteile, in welchen die Zeit eine Rolle spielt, immer Urteile über gleichzeitige Ereignisse sind." A. Einstein, "Zur Elektrodynamik bewegter Körper," *Annalen der Physik* 17 (1905): 891–921; *Collected Papers*, vol. 2, pp. 275–306. See also A. Einstein, "How I Created the Theory of Relativity," *Physics Today* 35 (August 1982): 45–47. For historical details about the concept of simultaneity, see also M. Jammer, "The History of the Concept of Distant Simultaneity," *Rendiconti delle Accademia Nazionale delle Scienze, Memorie di Scienze Fisiche* 103 (1985): 169–184.

At the beginning of the paragraph, Einstein defined the synchrony of two spatially separated clocks "of exactly the same constitution," say, clock C_A stationary at position A and clock C_B stationary at position B in an inertial coordinate frame, as follows. Suppose that at time t_A on clock C_A a light pulse leaves A toward B, arrives at B at time t_B on clock C_B and is immediately reflected to A, where it arrives at time t'_A on clock C_A. Then, provided $t_B - t_A = t'_A - t_B$, the clocks are defined as synchronized. Two spatially separated events e_A at A and e_B at B are defined as simultaneous—symbolically, $e_A \, \sigma \, e_B$—with reference to an inertial frame S, if synchronized clocks stationary at A and B show the same time at events e_A and e_B.

Einstein assumed, without proving this assumption, that simultaneity thus defined with respect to one and the same coordinate system is a symmetric and transitive relation; that is, that if $e_A \, \sigma \, e_B$, then $e_B \, \sigma \, e_A$, and that if $e_A \, \sigma \, e_B$ and $e_B \, \sigma \, e_C$, then $e_A \, \sigma \, e_C$, where e_C is an event at location C in the coordinate system.[14]

In the second paragraph of his 1905 paper, Einstein proved the relativity of simultaneity: "Two events that are simultaneous when observed from some particular coordinate system can no longer be considered simultaneous when observed from a system that is moving relative to that system." A simple demonstration of this relativity was later given by Einstein in his thought experiment of a long train, traveling along the rails on an embankment and hit by two lightning strokes "which are simultaneous with ref-

[14] For proofs of the symmetry and transitivity of the simultaneity relation see H. Reichenbach, *Axiomatik der relativistischen Raum-Zeit-Lehre* (Vieweg, Braunschweig, 1924); *Axiomatics of the Theory of Relativity* (University of California Press, Berkeley, 1969), pt. 1, sec. 7.

erence to the embankment but not simultaneous with respect to the train."[15]

Finally, that simultaneity is a reflexive relation—that is, that $e \sigma e$ holds for every event e—is generally accepted as intuitively evident even though, strictly speaking, not formally derivable from the definition of simultaneity because no event can be spatially separated from itself.

Having defined the concept of simultaneity, Einstein was able to define the "time" of an event in a given inertial reference frame as follows: "The time of an event is the reading obtained simultaneously with the event from a clock at rest that is located at the place of the event and that for all time determinations is in synchrony with a specified clock at rest."[16]

TURNING NOW from Einstein's definitions of simultaneity and time to theology, recall that the notion of time has always been of great significance for theological thought. This is not the place to retrace the history of the role of this notion in the Bible or in the theological contemplations of the early philosophers of ancient Greece. It suffices to recall that Plato, who was probably the first to use the term "theology," explained the existence of time within a theological context.[17] In the *Timaeus*, his allegorical exposition of the

[15] A. Einstein, *Über die spezielle und die allgemeine Relativitätstheorie* (Vieweg, Braunschweig, 1917); *Relativity—The Special and the General Theory* (Methuen, London, 1920), sec. 9.

[16] "Die 'Zeit' eines Ereignisses ist die mit dem Ereignis gleichzeitige Angabe einer am Orte des Ereignisses befindlichen, ruhenden Uhr, welche mit einer bestimmten ruhenden Uhr, und zwar für alle Zeitbestimmungen mit der nämlichen Uhr, synchron läuft." A. Einstein, "Zur Elektrodynamik . . . ," 1905, German original, p. 894; English trans., p. 143.

[17] Plato, *Politeia, Republic*, II, 379a.

drama of creation, Plato described how God—or the Demiurge—who is eternal, sought to create a universe that is also eternal; but "to attach Eternity in its entirety to what is generated was impossible; wherefore He resolved to make a moving likeness of Eternity which abides in unity. He made an eternal image, moving according to number, that which we name 'time.'[18] Plotinus similarly conceived of time as an image of eternity that he regarded as a changeless and timeless life of God or of the World Soul. "The life, then, which belongs to that which exists and is in being, all together and full [zoē homou pasa kai pleres], completely without extension or interval [adiastatos], is that which we are looking for, eternity."[19]

Early medieval theologians, like Saint Gregory of Nazianzus or Aurelius Augustinus (Saint Augustine) were deeply influenced by these Platonic ideas and declared that God created the world not *in* time but *together with* time ("non in tempore sed cum tempore finxit Deus mundum"). "It is not in time that you [God] precede times; all your 'years' subsist in simultaneity, because they do not change; your 'years' are 'one day' and your today is eternity," wrote Augustine in his *Confessions*.[20]

Because by implication, God's knowledge of temporal events is also timeless, the doctrine of divine timelessness seemed to resolve the problem of the incompatibility of God's foreknowledge and human free will. Although the history of this incompatibility problem can be traced back

[18] Plato, *Timaeus*, 37C–38A.

[19] Plotinus, *Enneads* III: 7, 11; 7, 18.

[20] Augustine, *Confessiones*, book 11; *Confessions* (Oxford University Press, Oxford, 1991), p. 230. This statement is followed by his often quoted question "Quid est tempus?" "What is time? If no one asks me, I know what it is. If I wish to explain it to him who asks me, I do not know."

to Marcus Tullius Cicero, it was St. Augustine's formulation that brought it to the attention of theologians.[21] He wrote, "Your trouble is this. You wonder how it can be that these two propositions are not contrary and incompatible, namely that God has foreknowledge of all future events and that we sin voluntarily and not by necessity. For if, you say, God foreknows that a man will sin, he must necessarily sin. But if there is necessity, there is no voluntary choice of sinning but rather fixed and unavoidable necessity."[22]

The most influential successor of Augustine in the quest of resolving the incompatibility problem was undoubtedly the sixth-century philosopher and theologian Anicius Manlius Torquatus Severinus Boethius. In his *Consolatio Philosophiae,* Boethius, obviously following Plotinus, defined eternity as follows: "Aeternitas est interminabilis vitae tota simul et, perfecta possessio."[23] And in his *De Trinitate,* he declared "Sempiternitas et aeternitas differunt. Nunc enim stans et permanens aeternitatem facit; nunc currens in tempore sempiternitatem."[24]

[21] M. T. Cicero, *De Divinatione,* books 2, 3.

[22] Augustine, *De Libro Arbitrio,* book 3, chap. 3, sec. 6. J. H. S. Burleigh, ed., *Augustine's Earlier Writings* (Westminster Press, Philadelphia, 1955), p. 173. For Augustine's attempt to solve this problem, see W. L. Rowe, "Foreknowledge and Free Will," *Review of Metaphysics* 18 (1964): 356–363.

[23] "Eternity is the complete possession all at once of illimitable life." Boethius, *The Theological Tractates and the Consolation of Philosophy* (Heinemann, London; Harvard University Press, Cambridge, Mass., 1973), p. 422.

[24] "There is a difference between the present of our affairs, which is now, and the divine present: our 'now' connotes changing time and sempiternity; but God's 'now,' abiding unmoved, and immovable, connotes eternity." Ibid.

Boethius' expression "tota simul" and his differentiation between eternity and temporality have been favorably discussed by "the great doctor of the Church" Saint Thomas Aquinas in his influential *Summa Theologica*, which opens by quoting Boethius and is widely accepted in Christian theology.[25] In short, Boethius' definition of eternity became the starting point of many theological doctrines of eternity and, even when criticized or rejected, remains so today, as shown in the writings of Paul Tillich, Karl Barth, and Oscar Cullman.

Most criticisms of Boethius' interpretation of divine eternity as "complete possession all at once [tota simul] of illimitable life" focus on the expression "tota simul" or that every "nunc currens" is simultaneous with the divine "nunc stans." Richard Swinburne, for example, pointed out that according to Boethius, God is

> simultaneously present at (and a witness of) what I did yesterday, what I am doing today, and what I will do tomorrow. But if t_1 is simultaneous with t_2 and t_2 with t_3, then t_1 is simultaneous with t_3. So if the instant at which God knows these things were simultaneous with both yesterday, today, and tomorrow,

[25] Thomas Aquinas, *Summa Theologica* (Great Books of the Western World, Encyclopedia Britannica, Chicago, Ill., 1952), vol. 19; see, e.g., question 14, article 13. Similar views of time and divine eternity also can be found in medieval and modern Jewish thought. Examples are: Crescas, *Or Adonai* I, 2, 11; Albo, *Ikkarim* II, 18–19; Maimonides, *Mishne Torah*, Tshuvot 5, 1; Bahya ibn Asher, *Commentary to the Pentateuch*, vol. 2, p. 134. Also according to the *Cabala* "God is timeless" ("lemaalah min ha-zeman"), see, e.g., J. Oshlag, *Tilmod Eser Sefirot*, vol. 1 pt. 1, p. 26; similarly, the eighteenth-century Hassidic Rabbi Nahman of Bretzlav wrote in his *Likkute Maharan* that "God is above time."

> then these days would be simultaneous with each other. So yesterday would be the same day as today and as tomorrow—which is clearly nonsense.[26]

In the same vein, Anthony Kenny wrote, "My typing of this paper is simultaneous with the whole of eternity. Again, on this view, the great fire of Rome is simultaneous with the whole of eternity. Therefore, while I type these very words, Nero fiddles heartlessly on."[27]

As explicitly stated—for example, by Swinburne—such objections are based on the transitivity of the simultaneity relation. This transitivity holds for the notion of simultaneity as used in common everyday language or as critically expounded by Leibniz and Kant or even as defined by Einstein with respect to a single reference frame.[28] It does not hold, however, if relativized to different reference frames, as Einstein has shown in his proofs of the relativity of simultaneity. It was precisely the Einsteinian relativity of simultaneity by which Eleonore Stump, then of Virginia State University, and Norman Kretzmann of Cornell University proposed in 1981 to eliminate the difficulties with the Boethian definition of eternity.[29] They noticed that, to use Kenny's example, in each of the simultaneity relations in the two premises, namely in the simultaneity of the ac-

[26] R. Swinburne, *The Coherence of Theism* (Oxford University Press, Oxford, 1977, 1986), pp. 220–221.

[27] A. Kenny, *The God of the Philosophers* (Oxford University Press, Oxford, 1979), pp. 38–39.

[28] See M. Jammer, "The History of the Concept of Distant Simultaneity," *Rendiconti* 103, p. 175.

[29] E. Stump and N. Kretzmann, "Eternity," *The Journal of Philosophy* 78 (1981): 429–458; reprinted in T. Morris, ed., *The Concept of God* (Oxford University Press, Oxford, 1987), pp. 219–252.

tion of typing with the whole of eternity and in the simultaneity of Rome's great fire with the whole of eternity, one of the two relata (i.e., the terms between which the relation holds) is temporal and the other is atemporal or eternal, whereas in the simultaneity relation in the conclusion, namely the simultaneity of the typing and Nero's fiddling during Rome's fire, both of the relata are temporal. They proposed therefore to distinguish three different types of simultaneity: (1) T-simultaneity, the existence or occurrence at one and the same time; (2) E-simultaneity, the existence or occurrence at one and the same eternal present; and (3) ET-simultaneity, which holds between what is eternal and what is temporal.[30]

For the sake of brevity and in accordance with the above, the ET-simultaneity relation is denoted σ_{ET}. Because one of the relata of σ_{ET} is eternal, the definition of σ_{ET} must refer to one and the same present and not to one and the same time. Because σ_{ET} deals with two equally real modes of existence, neither of which can be reduced to the other, "the definition must be constructed in terms of *two* reference frames and *two* observers." This, then, is the Stump and Kretzmann definition of σ_{ET}: Any two entities or events x and y satisfy $x \, \sigma_{ET} \, y$ if and only if

(1) either x is eternal and y is temporal, or vice versa; and

(2) for some observer, A, in the unique eternal refer-

[30] Quentin Smith's "A new typology of temporal and atemporal permanence," *Nous* 23 (1989): 307–330, which differentiates strictly among a multitude of types of duration (sempiternality, omnitemporality, everlastingness, eternalness, etc.), would make it possible to greatly increase the number of simultaneities. But Smith does not accept the ET-simultaneity proposed by Stump and Kretzmann.

ence frame, x and y are both present, that is, either x is eternally present and y is observed as temporally present, or vice versa; and

(3) for some observer, B, in one of the infinitely many temporal reference frames, x and y are both present, that is, either x is observed as eternally present and y is temporally present, or vice versa.

It follows from this definition that $x\ \sigma_{ET}\ y$ entails that x is neither earlier nor later than y and that x and y cannot both exist at one and the same time within a given observer's reference frame because one of them must be eternal. The relation σ_{ET} is symmetric, but it is not reflexive because no temporal or eternal x can, by definition, satisfy $x\ \sigma_{ET} x$. It is also not transitive, for if it were, its symmetry would imply its reflexivity. It is clear that the interpretation of the "simul" in "tota simul" as an ET-simultaneity immunizes, by virtue of its intransivity, Boethius' definition of eternity, at least against attacks such as those mentioned above. In fact, "from a temporal standpoint, the present is ET-simultaneous with the whole infinite extent of an eternal entity's life. From the standpoint of eternity, every time is present, co-occurrent with the whole of infinite atemporal duration."[31]

Stump and Kretzmann's essay on ET-simultaneity, which, as they acknowledge, was inspired by Einstein's concept of the relativity of simultaneity, caused quite a stir among theologians and philosophers. It was hailed by some as an admirable theoretical achievement, whereas others criticized it as being based on unintuitive conceptions. In reply to such charges, Stump and Kretzmann proposed an

[31] Stump and Kretzmann, "Eternity."

optical model in which light represents eternity.[32] It consists of two infinite horizontal lines, one on top of the other; the upper one is all along its length a strip of light and therefore represents eternity, whereas the lower line is completely dark except for a dot of light that moves steadily along it and represents a moment of time. "The light, in the representations of eternity and of time, should be interpreted as an indivisible present. For any instant of time as that instant is present, the whole of eternity is present at once; the infinitely enduring, indivisible eternal present is simultaneously with each temporal instant as it is the present instant."[33]

Optical analogues or models that explain the relation between divine eternity and mundane time had been used before. Thomas Aquinas had written, "All things that are in time are present to God from eternity, not only because He has the type of things present within Him, as some say, but because His glance is carried from eternity over all things as they are in their presentness."[34] The optical "seeing" was also used in the same context by Swinburne who wrote: "The obvious analogy is to men travelling along a road; at each time they can see only the neighbourhood on the road where they are. But God is above the road and can see the whole road at once. Taking man's progress along the road as his progress through time, the analogy suggests that while man can enjoy only one time at once, God can enjoy

[32] Recently Stump and Kretzmann expressed some dissatisfaction with their model. See B. J. Shanley, "Eternity and Duration in Aquinas," *Thomist* 61 (1997): 525–548.

[33] E. Stump and N. Kretzmann, "Atemporal duration," *Journal of Philosophy* 84 (1987): 214–219.

[34] Thomas Aquinas, *Summa Theologica.*

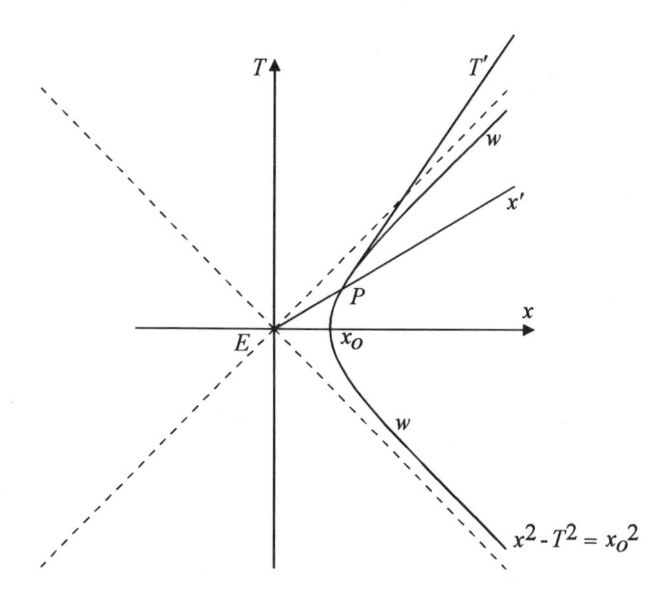

Fig. 1. Two-dimensional space-time representation of eternity.

all times at once."[35] A full-page diagram of such a procession along a road, representing all the main events in a person's life (and afterlife), has been given by Ralph G. Mitchell in his book *Einstein and Christ*.[36]

Theologians and philosophers do not seem to know that the special theory of relativity itself, by means of its space-time geometrical diagrams, offers perhaps the best representation of the eternity-time relation.

For the sake of simplicity, we confine our discussion to a two-dimensional space-time representation (Fig. 1) in

[35] R. Swinburne, *The Coherence of Theism*, pp. 216–217.

[36] R. G. Mitchell, *Einstein and Christ* (Scottish Academic Press, Edinburgh, 1987), p. 41.

which x denotes the spatial location of an event e and t the time of its occurrence with respect to a reference frame S. The space and time coordinates of e are x' and t' in another reference frame S', which is moving relative to S. The transformations that relate (x, t) to (x', t') are the Lorentz transformations, which Einstein derived in the third paragraph of his famous 1905 paper, "On the Electrodynamics of Moving Bodies." We use $T = ct$, where c denotes the velocity of light in a vacuum. The physical system in which the events occur in temporal sequence is assumed to move in S along a hyperbolic world line w defined by $x = (x_o^2 + T^2)^{\frac{1}{2}}$, where x_o is a constant. The space-time coordinates of an event P in the system are denoted with respect to S by (x_p, T_P) and its velocity by v_p. Hence

$$v_P = \frac{dx}{dt_P} = c\,\frac{dx}{dT_P} = c\,\frac{T_P}{(x_o^2 + T_P^2)^{\frac{1}{2}}} = c\,\frac{T_P}{x_P}. \tag{1}$$

This shows that v_P varies along w but is always less than c. Because the motion of the system is accelerated, a local inertial reference frame S_P must be introduced with P as its origin, that is, an inertial frame (x', T') in which the system is momentarily at rest in P, so that $x'_P = T'_P = 0$. The Lorentz transformation of the time coordinates is

$$T' = \gamma\left[(T - T_P) - \beta\,(x - x_P)\right], \tag{2}$$

where $\beta = v_P/c$ and $\gamma = (1 - \beta^2)^{-\frac{1}{2}}$. The x'-axis, on which $T' = 0$ always, therefore satisfies the equation

$$\frac{T - T_P}{x - x_P} = \frac{v_P}{c} = \frac{T_P}{x_P}, \tag{3}$$

where (1) has been used. But equation (3) is also satisfied by the origin E, where $x = T = 0$. It follows that E, the "eternity," is always simultaneous with the event P with respect to the local inertial frame. But P is an arbitrary event on world line w of the system. Hence, all events, which occur in the system in temporal sequence, are simultaneous with E, or more precisely are ET′-simultaneous with E, in full accordance with the Boethian "tota simul." Temporally different events in the system are simultaneous with E without being simultaneous among themselves as is clearly explained by the fact that each P has its specific S_P so that the relativity of simultaneity abrogates its transitivity.

In contrast to the model of Stump and Kretzmann, in which eternity is represented as an infinitely long (illuminated) line, eternity is represented here by a point, that is, as something without spatial (or temporal) extension. This agrees very well with scholars who interpret the term "interminabilis" in the Boethian definition of eternity not as "infinite" but rather as "unextended" in the sense of a "point," which according to the very first definition in Euclid's *Elements* "has no part." In fact, Boethius described the relationship between time and eternity as that between a circle and its center.[37] Moreover, Plotinus had already conceived of eternity as a pointlike entity "without extension or interval," as quoted above, or as "a point [sēmeion] to which all lines converge."[38]

[37] "[U]t est . . . ad aeternitatem tempus, ad punctum medium circulus." Boethius, *Consolatio Philosophiae*, book 4, prose 6.

[38] Plotinus, *Enneads* III, 7, 3. See also K. A. Rogers, "Eternity Has No Duration," *Religious Studies* 30 (1994): 1–16.

In the recent theological literature, scarcely another essay attracted so much attention and received so many responses, for and against, as the Stump and Kretzmann essay "Eternity."[39] It is beyond the scope of this book to discuss all these criticisms in any detail, even though references to the theory of relativity play a decisive role in some of them. The essay "Eternity" also had the effect of bringing many theologians to study the fundamentals of the theory of relativity. Endorsing this new trend, William Lane Craig described the situation with these words:

[39] D. B. Burrell, "God's Eternity," *Faith and Philosophy* 1 (1984): 389–406. S. B. Cowan, "A Reductio ad Absurdum of Divine Temporality," *Religious Studies* 32 (1996): 371–378. W. L. Craig, "The Special Theory of Relativity and Theories of Divine Eternity," *Faith and Philosophy* 11 (1994): 19–37. P. Fitzgerald, "Stump and Kretzmann on Time and Eternity," *Journal of Philosophy* 82 (1985): 260–269. W. Hasker, "Concerning the Intelligibility of 'God is Timeless'" *New Scholasticism* 57 (1983): 170–195; *God, Time and Knowledge* (Cornell University Press, Ithaca, N.Y., 1989). B. Leftow, "The Roots of Eternity," *Religious Studies* 24 (1988): 189–212; "Time, Actuality and Omniscience," *Religious Studies* 26 (1990): 303–322; *Time and Eternity* (Cornell University Press, Ithaca, N.Y., 1991); "Eternity and Simultaneity," *Faith and Philosophy* 8 (1991): 148–179; D. Lewis, "Eternity Again: A Reply to Stump and Kretzmann," *International Journal for Philosophy of Religion* 15 (1984): 73–79; "Timelessness and Divine Agency," *Philosophy of Religion* 21 (1987): 148–159; "Eternity, Time and Tenselessness," *Faith and Philosophy* 5 (1988): 72–86; H. Nelson, "Time(s), eternity, and duration," *Philosophy of Religion* 22 (1987): 3–19; A. G. Padgett, *God, Eternity and the Nature of Time* (St. Martin's Press, New York, 1992), esp. pp. 66–74; "Eternity and the Special Theory of Relativity," *International Philosophical Quarterly* 33 (1993): 219–233; J. Simon, "Eternity, Omniscience and Temporal Passage: A Defence of Classical Theism," *Review of Metaphysics* 42 (1989): 547–568. See also E. Stump and N. Kretzmann, "Eternity, Awareness, and Action," *Faith and Philosophy* 9 (1992): 463–483, in which they reply to some of the criticisms raised against their 1981 paper.

> Current discussions of God's eternity have been for the most part carried out in almost complete ignorance of the philosophy of space and time and without any profound knowledge of Relativity Theory and its analysis of time—a remarkable shortcoming, when one thinks about it, for how can one pretend to formulate an adequate doctrine of God's eternity and His relationship to time without taking cognizance of what modern philosophy and science have to say about time?[40]

In particular, Craig cited a former student of Richard Swinburne at Oxford, Alan G. Padgett, who had demonstrated the fruitfulness of referring to the theory of relativity in theological discourse in his Ph.D. dissertation and later publications.[41]

Padgett criticized ET-simultaneity as being based on a false analogy. According to the special theory of relativity, space-time is divided, relative to an observer at the event *e* and apart from *e*, into four disjointed domains: the *absolute past*, which contains all events that are prior to *e* in every reference frame; the *absolute future*, which contains all events that are posterior to *e* in every reference frame; the *elsewhere* or *conditional present*, which contains all events that can be made simultaneous with *e* by an appropriate choice of the reference frame; and the *light cone* itself, which consists of all events that can be connected with *e* by a light ray. Furthermore, because the theory of relativity

[40] W. L. Craig, "God and Real Time," *Religious Studies* 26 (1990): 335–347.

[41] A. G. Padgett, "Divine Eternity and the Nature of Time," Ph.D. diss., Oxford University, 1988.

states that any causal interaction cannot spread faster than the velocity of light, any event that is causally connected with *e* must lie inside the light cone or on its surface, its apex being at *e*. Hence, two events that are outside of each other's light cones cannot be causally connected. "Since Stump and Kretzmann want some sort of causal connection between God and the world to hold, they should not have appealed to the relativity of simultaneity between events that—by the very nature of things—cannot be causally connected."[42]

Padgett also objected that a simultaneity between timeless eternity and human time would destroy the distinction between temporality and timelessness. "If there are timeless beings, one would think they could not, by definition, be 'at the same time as' (simultaneous with) something in time. How can something timeless and something in time be at the *same* time?" He concluded that "ET-simultaneity is *ex hypothesi*, self-contradictory."

As stated above, the idea of God's timeless eternity or of a timeless mode of divine existence has been conceived in order to resolve the problem of the incompatibility between divine omniscience (including foreknowledge) and human free will. It has also been argued, however, that divine omniscience is incompatible with timelessness. To cite a very simple argument, given by Arthur Prior: A timeless being cannot know that today is the Fourth of July.[43] The argument that a God who knows everything

[42] A. G. Padgett, *God, Eternity and the Nature of Time*, pp. 69–72.

[43] A. Prior, "The Formalities of Omniscience," *Philosophy* 37 (1962): 114–129. For other arguments against the compatibility between divine omniscience and timeless eternity, see, e.g., N. Pike, *God and Timelessness* (Routledge and Kegan Paul, London, 1970); W. Hasker,

must exist in time agrees, of course, with the Judeo-Christian religious tradition in which God acts, rewards, punishes, and accedes to petitionary prayers. But does divine temporality imply also divine spatiality? It has been argued that the special theory of relativity provides a positive answer to this question.

Assigning the attribute of spatiality to God's existence constitutes a problem because, with the exception of certain early Christological doctrines, God was generally conceived of as an immaterial, incorporeal being. This conception was not adequately expressed in the Scriptures; but as early as Philo of Alexandria and the Greek patristic theology—for example, with Gregory of Nyssa—the incorporeity of God (asomatotes) became a fundamental tenet of faith. Later, in the Middle Ages, the third of Maimonides' thirteen Articles of Faith declared: "I firmly believe that the Creator, blessed be His name, is not corporeal and no bodily attributes apply to Him." God was conceived as a purely intellectual, spiritual, or mental agent. The general problem was therefore whether such a mode of existence necessarily exists in space. Robert Weingard has given an affirmative answer, provided that the mental agent acts in time, that is, his actions are datable. Weingard argued that, according to Einstein's theory of relativity, all

> events belong to a spatiotemporal structure, space-time, in which spatial and temporal relations are intermixed and cannot be separated from each other. So for mental events to be part of the network of temporal relations that events bear to each other, they must be

God, Time, and Knowledge, chap. 8, 9; J. Simons, "Eternity, Omniscience and Temporal Passage," *Review of Metaphysics* 42 (1989): 547–568.

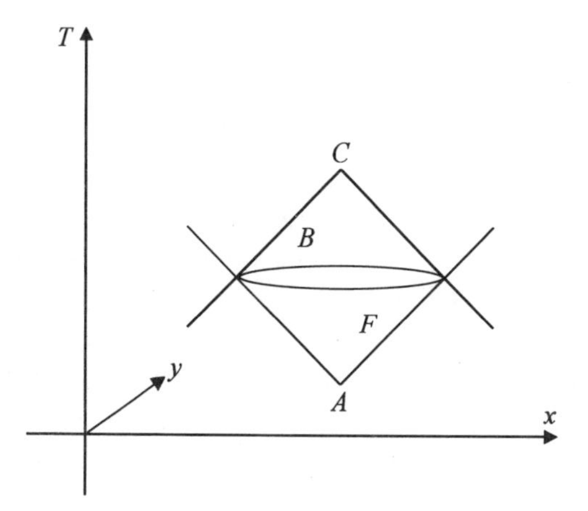

Fig. 2. The geometrical structure of the four-dimensional Minkowksi space-time. The third spatial dimension, corresponding to the *z*-axis is suppressed.

part of the space-time network of relations, for it is simply not plausible to suppose there is, in addition to space-time, a network of purely temporal relations for the mental events to belong to.[44]

An argument that is somewhat similar, but based on the additional premise that mind and body are capable of interacting with each other, has been presented by Michael Lockwood in terms of the geometrical structure of the four-dimensional Minkowski space-time (Fig. 2). Let A' be a physical event which is the cause of a mental event E that, in turn, is the cause of a physical event C'. Because they are physical events, A' and C' can be represented by space-time points A and C. Because the theory of rela-

[44] R. Weingard, "Relativity and the Spatiality of Mental Events," *Philosophical Studies* 31 (1977): 279–284.

tivity states that a causal action cannot travel faster than the speed of light, *E*, the effect of *A*, must lie, like every event caused by *A*, in *F*, the forward light cone of A. For the same reason, *E*, the cause of *C*, must lie, like every event capable of causing *C*, in the backward light cone *B* of *C*. It follows therefore that the mental event must lie in the intersection of the two light cones, a region that in principle can be made as small as desired. The mental event *E* must consequently have a definite spatial location. "If Einstein was right about space and time," Lockwood concluded, "mental events must belong to the same spatial order as physical events."[45]

ET-SIMULTANEITY, based on the relativity of simultaneity, was defined in order to save the classical conception of divine eternity; this conception, in turn, was devised by Boethius and others in order to defend the doctrine of free will or, as theologians call it, libertarianism, against the charge of its incompatibility with God's omniscience and foreknowledge. In short, the relativity of simultaneity has been applied to save libertarianism as a fundamental tenet of traditional religion and morality. In a much discussed article from 1966, Cornelis Willem Rietdijk claimed that, contrariwise, the relativity of simultaneity implies strict determinism and therefore necessitarianism, the denial of free will. Strictly speaking, this claim is much older, for it is part of the interpretation of space-time as a "block universe;" a term that was used as early as 1883 by the dialec-

[45] M. Lockwood, "Einstein and the Identity Theory," *Analysis* 44 (1984): 22–25.

tical metaphysician Francis Herbert Bradley to denote the detemporalization of physical reality.[46] Not only Hermann Weyl, whose characterization of the "block universe" was cited earlier, but also other philosophers and scientists, including Ernst Cassirer and, most eloquently, James Hopwood Jeans, expressed the idea that the theory of relativity implies strict determinism, the concept of the world as a "block universe," and the denial of free will, because clearly the Parmenidean doctrine that there is no "becoming" but only "being" requires that free will is at best an illusion.[47]

In spite of his belief in strict determinism, Einstein—with the exception of the letter written to Besso's family to palliate their grief—never explicitly expressed his adherence to a view such as the one propounded by Jeans. He emphatically declared in his "Credo" that he did not believe in the existence of free will.[48] According to C. W. Riet-

[46] F. H. Bradley, *The Principles of Logic* (Kegan Paul, London, 1883), p. 53.

[47] The original German edition of Weyl's book, *Philosophy of Mathematics and Natural Science*, was published in 1927 under the title *Philosophie der Mathematik und Naturwissenschaft* by Oldenburg in *Handbuch der Philosophie*.

E. Cassirer, *Zur Einsteinschen Relativitätstheorie* (B. Cassirer, Berlin, 1921), p. 119; *Einstein's Theory of Relativity* (bound together with *Substance and Function*), (Dover, New York, 1953), p. 449.

Sir James Jeans, "Man and the Universe" in *Scientific Progress* (George Allen and Unwin, London, 1936), pp. 11–38.

For criticisms of this view, see, e.g., P. Frank, "Is the Future Already Here?" *Interpretations and Misinterpretations in Modern Physics* (Hermann, Paris, 1938), pp. 46–55; reprinted in M. Capek, ed., *The Concepts of Space and Time* (Reidel, Dordrecht, 1976), pp. 383–395.

[48] F. Herneck, "Albert Einstein gesprochenes Glaubensbekenntnis."

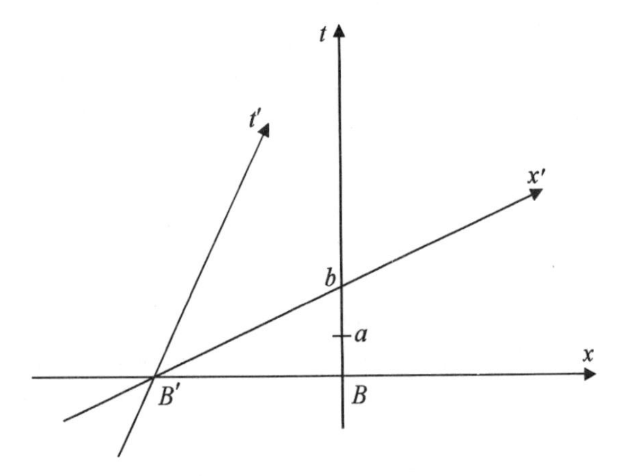

Fig. 3. Rietdijk's two inertial space-time coordinate systems (shown in two dimensions).

dijk, Einstein could have deduced this denial as a logical consequence of his theory of relativity particularly, the relativity of simultaneity.

Rietdijk considered the following physical scenario, which, for the sake of simplicity, is only two-dimensional. $S(x,t)$ and $S'(x',t')$ denote two inertial space-time coordinate systems, and B and B' are observers at their spatial origins, that is, at $x = 0$ and $x' = 0$, respectively; B' is moving toward B with constant velocity v (as measured in S), and the distance between B and B', when $t = 0$ and $t' = 0$, is d (see Fig. 3). The Lorentz transformation equations for this physical scenario are

$$x' = \gamma\,(x + d - vt) \tag{4}$$

and

$$t' = \gamma\left[t - (v/c^2)\,(x + d)\right], \tag{5}$$

where $\gamma = (1 - v^2/c^2)^{-\frac{1}{2}}$. Let a be an event with coordinates $x_a = 0$ and $t_a > 0$ so that it lies in the future of B at ($x = 0$; $t = 0$). According to (5), its time coordinate in S' is

$$t'_a = \gamma \left[t_a - (v/c^2)\, d\right], \tag{6}$$

which shows that, for every such t_a and every d, velocity v can be chosen so that $t'_a < 0$. This means that event a lies in the past of B' at ($x' = 0$; $t' = 0$) and is therefore, according to Rietdijk, already "given" and "determined" for B' at ($x' = 0$; $t' = 0$). Because B' at ($x' = 0$; $t' = 0$) is in the present or "now" of, and simultaneous with, B at ($x = 0$; $t = 0$), Rietdijk concluded that event a is determined also for B at ($x = 0$; $t = 0$), at least in the sense that B cannot possibly influence event a, for example by preventing its occurrence. But a is an arbitrary event in the future of B. Hence, "each event is determined; it is already past for someone in our 'now,'" a result that obviously excludes the possibility of free will.[49] Rietdijk's argument is based on the fact that, for any two events e_1 and e_2 in relativistic space-time, there exists a third event, e_3, in spacelike relation to both of them. This can easily be seen by considering the topologically closed interiors of the two light cones, or rather hypercones, whose apexes are at e_1 and e_2. That the argument is also based on the relativity of simultaneity follows at once if we choose, instead of event a, event b, defined by the intersection of the t-axis with the x'-axis. Obviously, b is simultaneous with B' at ($x' = 0$; $t' = 0$), and B' is simultaneous with B at ($x = 0$; $t = 0$), but b is *ex hypothesi* in the future of B at ($x = 0$; $t = 0$) and therefore not simul-

[49] C. W. Rietdijk, "A Rigorous Proof of Determinism Derived from the Special Theory of Relativity," *Philosophy of Science* 33 (1966): 341–344.

taneous with it. This nontransitivity verifies the relativity of the simultaneity relation with respect to different reference frames as it is used in the argument.

In the introduction to his 1971 book, Rietdijk wrote that "up to the present, neither any refutation of that article [his 1966 paper], nor even an attempt thereto has been published. I think, therefore, that we may consider determinism is established, with all the important consequences of it. . . . Also there is no free will."[50] Rietdijk seems to have been unaware that Howard Stein in 1968 and Peter Theodore Landsberg in 1970 independently severely criticized the logic of his arguments.[51] More recently, Roberto Torretti has given a very lucid critical analysis of this chronogeometrical substantiation of strict determinism or necessitarianism.[52]

It should be noted in this context that the well-known cosmologist Hermann Bondi claimed in 1952 that "relativity demands a non-deterministic theory such as is given at present by quantum theory." He summarized his argument as follows, "The flow of time has no significance in the logically

[50] C. W. Rietdijk, *On Waves, Particles and Hidden Variables* (Van Gorcum, Assen, The Netherlands, 1971), p. IV.

[51] H. Stein, "The Einstein-Minkowski Space-Time," *The Journal of Philosophy* 65 (1968): 5–23.

P. T. Landsberg, "Time in Statistical Physics and Special Relativity," *Studium Generale* 23 (1970): 1108–1158; reprinted in J. T. Fraser, F. C. Haber, and G. H. Muller, eds., *The Study of Time* (Springer, Berlin, 1972), pp 59–109. A reasoning similar to that of Rietdijk's and subject to the same criticisms is H. Putnam's argument in his paper "Time and Physical Geometry," *Journal of Philosophy* 64 (1967): 240–247. For a critique of Putnam's argument, see also L. Sklar, *Philosophy and Spacetime Physics* (University of California Press, Berkeley, 1985), pp. 290–292.

[52] R. Torretti, *Relativity and Geometry* (Pergamon Press, Oxford, 1983), pp. 249–251.

fixed pattern of events demanded by deterministic theory, time being a mere co-ordinate. In a theory with indeterminacy, however, the passage of time transforms statistical expectation into real events."[53] Bondi's argument, unlike Rietdijk's opposing claim, seems not to have gained any attention. Nor does Einstein seem to have taken any notice of it.

WERE RIETDIJK'S disproof of the impossibility of free will correct, it would eliminate, of course, the incompatibility problem of divine omniscience and libertarianism because free will would not exist. Another solution of the incompatibility problem could be obtained if God's infallible omniscience were restricted to knowledge of only past and present events; then the problem could not be raised because, to use Augustine's formulation of it, God could not foreknow that man would sin.[54]

A theological doctrine that professes precisely this limited conception of divine omniscience is "process theology," sometimes also called "new theology," which originated as a movement of reorientation in American Protestant thought after the Second World War and subsequently gained some attention in Europe. This doctrine was influenced by Alfred North Whitehead's metaphysics as presented, for example, in his book *Process and Reality*, which replaces the mechanistic physical world picture of substances by an organic conception of reality as a network of events in space and time.[55] Process theology contends that objects of knowledge are not fully determined as long as they are

[53] H. Bondi, "Relativity and Indeterminacy," *Nature* 169 (1952): 660.

[54] Augustine, *De Libro Arbitrio*, book 3, chap. 3.

[55] A. N. Whitehead, *Process and Reality* (Macmillan, New York, 1929; Free Press, New York, 1978).

future events. Accordingly, even though God knows the present probability of the actualization of a future event, he does not know in advance whether the event will be actualized or not. God's ignorance of future events is, however, not an imperfection of his knowledge because there is nothing in the future for anyone to be ignorant of.

As a matter of fact, such a limitation of divine knowledge had already been considered, though eventually rejected, by Thomas Aquinas when he wrote in his *Summa Theologica*, "Everything known by God must necessarily be, because even what we ourselves know must necessarily be; and, of course, the knowledge of God is much more certain than ours. But no future contingent thing must necessarily be. Therefore no contingent future thing is known to God."[56]

The implications of Whitehead's philosophy for theology have been elaborated primarily by Charles Hartshorne in his writings on "temporalistic theism."[57] Because all realities are processes of realizations and because God and the world condition each other, God's knowledge, which encompasses all realities, but only realities, grows and increases in an immanent way with the process of their development. God both participates in, and is enriched by, the processes of the natural order. God is not a transcen-

[56] "Omne scitum a Deo necesse est esse, quia etiam omne scitum a nobis necesse est esse; cum tamen scientia Dei certior sit, quam scientia nostra; sed nullum contingens futurum necesse est esse; ergo nullum contingens futurum est scitum a Deo." *Divi Thomae Aquinatis Summa Theologica*, I, question 14, article 13 (Senatus, Rome, 1886), p. 141.

[57] C. Hartshorne, *Whitehead's Philosophy—Selected Essays 1935–1970* (University of Nebraska Press, Lincoln, 1972; see esp. the essay "Whitehead's Idea of God," pp. 63–97.

dental atemporal being "above" the world. Rather, according to process theology, which identifies "being" with "becoming," God is immanent in the world, just as the world is immanent in God, although God and the world do not form an identity.[58] These ideas led Hartshorne to raise the question, "Must there not be a cosmic present, in spite of relativity physics, the de facto totality of actual entities as present in the divine immediacy?" In support of an affirmative answer, Hartshorne referred to philosophers like Henri Bergson, who contended that the inability of physicists to determine, by signaling methods, a unique cosmic present or an absolute—that is, nonrelativistic—simultaneity, need not prevent God from experiencing such a present, because God knows all events directly by intuiting them wherever they occur.

William A. Christian challenged Hartshorne's claim that the theory of relativity was not competent to deal with this theological issue. Christian argued that Whitehead's God need not effect in himself and in the world a mutual immanence of simultaneous events so that the relativity of simultaneity does not apply.[59]

Having studied the relation between divine omniscience and relativistic simultaneity, John T. Wilcox questioned the logical cogency of Christian's rebuttal of Hartshorne's claim. Wilcox argued that the theory of relativity poses serious problems not only to temporalistic theism but to any theism that conceives God as "an experiencing subject, possessing something like thought or consciousness, as a

[58] Cf. J. Cobb and D. Griffin, *Process Theology: An Introductory Exposition* (Westminster Press, Philadelphia, 1976).

[59] W. A. Christian, *An Interpretation of Whitehead's Metaphysics* (Yale University Press, New Haven, Conn., 1959), pp. 330–333.

knower of temporal processes . . . and having knowledge which is itself subject to growth."[60] Wilcox also pointed out that the thesis that God's knowledge increases and changes with the temporal sequence of events "seems to be compatible with the Hebrew notion of a God actively involved in history."[61]

The problem which process theology faces, Wilcox declared, is the incompatibility of omniscience not with free will but with the theory of relativity or as he phrased it: "How there can be divine knowledge of a relativistic universe?"

In order to understand the problem, Wilcox considered the following physical scenario (Fig. 4). Let e denote an event that occurs at location A and e_{-3}, e_{-2}, e_{-1}, e_0, e_1, e_2, e_3 a sequence of events that occurs at location B in an inertial reference frame S. Suppose e_{-3} lies on the backward light cone and e_3 on the forward light cone whose apex is at e. Hence, all events in the sequence between e_{-3} and e_3 are spacelike related to e and can therefore, according to the relativity of simultaneity, be made simultaneous with e by

[60] J. T. Wilcox, *Relativity, Simultaneity, and Divine Omniscience* (M. A. thesis, Emory University, Atlanta, Ga., 1956).

[61] J. T. Wilcox, "A Question from Physics for Certain Theists," *Journal of Religion* 41 (1961): 293–300. In support of his last mentioned remark, Wilcox could have referred to *Genesis* (18: 32) in the Old Testament where it is reported how Abraham implored God not to destroy Sodom for the sake of its righteous inhabitants and God agreed, saying "I will not destroy it for the sake of ten," or to Exodus (32: 14) where it is said that Moses pleaded with God after the episode with the Golden Calf and "the Lord relented from the harm which He said He would do to His people." And, perhaps most importantly, that God has no foreknowledge of the consequences of even his own actions is asserted in *Genesis* (6: 6), "And it repented the Lord that he had made man on the earth, and it grieved him at his heart."

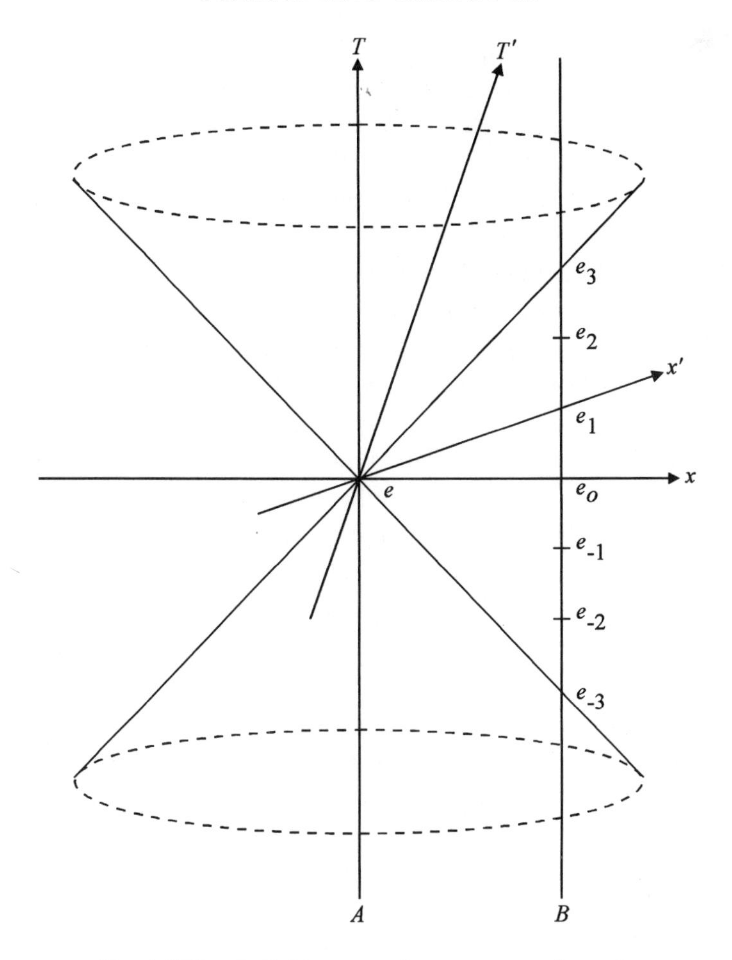

Fig. 4. Wilcox's light cone and a sequence of events.

an appropriate choice of reference frame. For example, e_1 is simultaneous with e in the reference frame with the axes x' and T'. But in a temporalistic theology, like process theology, one and only one event at B in the open interval between e_{-3} and e_3 can and must be perceived by God as being simultaneous with e. God's perception of simultaneous events would thus distinguish one special reference frame from all other possible frames and assign to it a privileged status in contradiction to Einstein's theory of relativity. Wilcox concluded therefore that the theory of relativity "throws a great deal of doubt upon the theory of temporalistic theism," and he asked, "Can one who accepts Einsteinian relativity believe in temporalistic theism?"[62]

The problem is even more complicated than Wilcox realized. He seems to have been unfamiliar with the thesis of the conventionality of intrasystemic simultaneity. According to this thesis, the truth of the statement that an event at B between e_{-3} and e_3 is simultaneous with the event e at A within a single inertial reference frame depends on the value of a real number ϵ between 0 and 1. This follows from the possibility of generalizing Einstein's synchronization condition, $t_B - t_A = t'_A - t_B$, or equivalently, $t_B = t_A + \frac{1}{2}(t'_A - t_A)$, to $t_B = t_A + \epsilon\,(t'_A - t_A)$, where $|\,\epsilon\,| < 1$, without ever running into conflict with physical experience. It is easy to see that for $\epsilon \neq \frac{1}{2}$ the one-way velocity of light in a given direction differs from that in the opposite direction, and both velocities differ from c. A process theologian would therefore have to say that God's perception of simultaneity would not only determine a preferred reference frame but also the numerical value of the velocity of light as measured in a given direction.[63]

[62] Ibid., p. 296.

[63] For details of the conventionality thesis of intrasystemic simultaneity see H. Reichenbach, *Philosophie der Raum-Zeit-Lehre* (Walter de Gruyter, Berlin, 1928); *The Philosophy of Space and Time* (Dover, New York, 1958), chap. 2; A. Grünbaum, *Philosophical Problems of Space and Time* (Knopf, New York, 1962; Routledge and Kegan Paul, London, 1964; Reidel, Dordrecht, 1973), chap. 12. An elementary introduction

Considering only the intersystemic relativity of simultaneity, Wilcox suggested the following solution for the problem:

> The assumption of a divine simultaneity need not mean that some actual perspective in the world is "right" as against others. For the divine perspective might be "eclectic," agreeing (approximately) as to some items with one standpoint, as to others with another, and the incidence of agreement might be constantly shifting. The number of moving bodies (at least in a finite area) being finite, they constitute all together only a vanishing fraction of the possible standpoints. There is then every reason to attribute a unique standpoint, not represented in the world, to deity, and in this way the "symmetry" among inner-worldly standpoints which is affirmed by physicists may be essentially retained.[64]

Wilcox's solution was rejected by Lewis S. Ford on the grounds that the assumption that God's perspective is eclectic and shifts constantly in perceiving the physical universe from one actual reference frame to another would create strange anomalies. It would imply the existence of events that would never be perceived even though they are theoretically accessible to perception. Moreover, "if

---

for the nonphysicist is J. Saunders and J. Norton, "Einstein, Light Signals and the ϵ-decision," in R. McLaughlin, ed., *What? Where? When? Why?* (Reidel, Dordrecht, 1982), pp. 101–127. But see also D. Malament, "Causal Theories of Time and the Conventionality of Simultaneity," *Nous* 11 (1977): 293–300; and A. Janis, "Simultaneity and Conventionality" in R. S. Cohen and L. Laudin, eds., *Physics, Philosophy and Psychoanalysis* (Reidel, Dordrecht, 1983), pp. 101–110.

[64] J. T. Wilcox, *Relativity, Simultaneity and Divine Omniscience.*

more radical shifts occurred, it would be possible for God to have the event he had just experienced as already past still before him in his future." Ford proposed to resolve the problem by postulating that although God's prehension may depend on space-time locality, it is not associated with a particular inertial reference frame so that relativistic effects would be completely avoided.[65]

The theological implications of the theory of relativity, which we have discussed so far, referred to the simultaneity of spatially separated events and to the relativity of this simultaneity, in accordance with the first paragraph of Einstein's 1905 paper on relativity. We now turn from the relativity of simultaneity to the relativity of time or time dilation and its theological applications. The physical theory relevant to this topic has been developed by Einstein in the subsequent paragraphs of this paper. In the third paragraph, he derived from the fundamental postulates—the relativity principle and the light postulate—the so-called Lorentz transformations. He thus obtained for the transformation of the time coordinates the equation,

$$t' = \gamma (t - vx/c^2), \tag{7}$$

where t is the time coordinate in an inertial reference frame $S(x,t)$ and t' that in a frame $S'(x',t')$ whose x'-axis is moving with constant velocity v along the x-axis of S, and the origins of the two frames coincide at $t = t' = 0$. As in equation (2), $\gamma = (1 - v^2/c^2)^{-\frac{1}{2}}$. The location of a clock stationary at the origin of S' is obviously in S given by

$$x = vt. \tag{8}$$

[65] L. S. Ford, "Is Process Theism Compatible with Relativity Theory?" *Journal of Religion* 48 (1968): 124–135.

Hence,

$$t' = \gamma (t - v^2t/c^2) = \gamma^{-1} t = t - (1 - \gamma^{-1}) t, \quad (9)$$

where t' is the elapsed time registered by the clock stationary in S' from the moment when it was at the origin of S until it arrives at x. Einstein concluded therefore that the time registered by the clock stationary in S' is slow relative to the clocks stationary in S by $1 - \gamma^{-1}$ seconds per second or, neglecting magnitudes of fourth and higher order, by $\frac{1}{2}v^2/c^2$ seconds per second. In other words, the "proper time" of a moving observer, that is, the time measured by him with a co-moving clock, lags behind the time measured by clocks stationary in the frame relative to which he is moving. This relativistic effect is called "time dilation." The ensuing result that the time difference between two events is frame-dependent is referred to as "the relativity of time." It is one of the philosophically most important results of the theory. It has also been widely popularized by the thought experiment of a traveler at high speed through space who, upon return, finds the world aged very much more than himself. Although such a journey through space is only a figment in science fiction, the time dilation effect is incontestably confirmed by experimental techniques in optics (e.g., by the transverse Doppler effect), in elementary particle physics (experiments with decaying particles), and in nuclear physics (by almost every experiment with cyclotrons).

The relativity of simultaneity is, of course, merely a special case of the relativity of time. If e_1 and e_2 are two spatially separated simultaneous events in the reference frame $S(x,t)$, so that the space-time coordinates of e_1 are (x_1,t_1) and those of e_2 are (x_2,t_2), where $x_1 - x_2 \neq 0$ but $t_1 = t_2$, then by equation (7),

$$t'_1 = \gamma\,(t_1 - vx_1/c^2) \quad \text{and} \quad t'_2 = \gamma\,(t_1 - vx_2/c^2), \tag{10}$$

and, by substraction,

$$t'_2 - t'_1 = \gamma\, v\,(x_1 - x_2)/c^2, \tag{11}$$

which differs from zero for $v \neq 0$. Hence, e_1 and e_2 are not simultaneous with respect to the reference frame $S'(x',t')$. From a strictly logical point of view, where the general case precedes the particular case, we should have dealt with the theological implications of the relativity of time and time dilation before our discussion of the implications of the relativity of simultaneity. Only the fact that we agreed to follow the order of Einstein's 1905 paper, which treats the relativity of simultaneity (or clock synchronism) in the second paragraph and the dilatation of time in the fourth paragraph, justifies our procedure.

IN ORDER to reconcile revelation and reason or religion and science, the notion of dilation of time has repeatedly been used by those for whom the Bible is not an allegorical or mythological but a true and accurate account of historical events. They firmly believe that, in particular, the biblical account of the creation of the world, as reported in the first chapter of Genesis, does not conflict, if correctly interpreted, with the findings of modern archeology, paleontology, and cosmology. Scripture says that God created the world, the heavens, and mankind in the course of six days. According to Hebrew tradition, God's creation of the world occurred less than six millennia ago. In the seventeenth century, James Ussher, archbishop of Armagh and primate of Ireland, compiled a chronology of the world, according to which, as stated in a margin to the King James Version of the Bible, creation occurred in 4004 B.C. This result was soon improved by the Cambridge theologian John Lightfoot to October 23, 4004 B.C. at 9 A.M. But totally unrelated modern scientific methods like the measurement of the Hubble constant and radioactive dating lead to the inescapable conclusion that the age of the universe is at least on the order of 15 billion years.

Although, strictly speaking, the proposal to resolve this conflict assumes some knowledge of the physical conditions of the early universe and its expansion, it suffices for our present issue, to understand that because of the expansion, different regions of the universe move with different velocities and thus represent local reference frames in relative motion to each other so that the conditions for the application of time dilation are locally satisfied.

The most detailed exposition of the relativistic solution of the chronological conflict is probably the second chapter, "Stretching of Time," in Gerald Schroeder's *Genesis and the Big Bang*.[66] Two quotations at the beginning of this chapter epitomize its contents. The first quotation is Maimonides's statement, "Conflicts between science and religion result from misinterpretations of the Bible," and the other is Einstein's equation "$t'_2 - t'_1 = \gamma^{-1} (t_2 - t_1)$," which follows from equation (9) above. In fact, Schroeder claimed that the correct interpretation of the biblical account of creation dispels any disagreement with modern physics. In order to explain how "to stretch six days to encompass 15 billion years" or "how to squeeze 15 billion years into six days," Schroeder quotes Psalm 90:4: "A thousand years in your

[66] G. Schroeder, *Genesis and the Big Bang* (Bantam Books, New York, 1990).

eyes are as a day that passes." "This verse," he continued, "is reminiscent of the dilation of time dealt with in Einstein's revolutionary thought experiments. Einstein demonstrated that when a single event is viewed from two frames of reference, a thousand or even billion years in one can indeed pass for days in the other."[67] Schroeder therefore concluded,

> In the first days of our universe's existence, the Eternal clock saw 144 hours pass. We now know that this quantity of time need not bear similarity to the time lapse measured at another part of the universe. As dwellers within the universe, we estimate the passage of time with clocks found in our particular, local reference frame; clocks such as radioactive dating, geologic placement, and measurements of rates and distances in an expanding universe. It is with these clocks that humanity travels.[68]

In order to resolve the conflict between the biblical statement that the creation of the world lasted only six days and the geological or paleontological estimates of billions of years for the age of the earth, Cyril Domb, a leading expert on statistical mechanics, resorted, like Schroeder, to the relativistic time dilation effect. The simplest solution, wrote Domb, is to assume

> that the word "day" in the six days of creation is to be interpreted as a period of time, an idea anticipated al-

[67] Ibid., p. 34. The expression "a single event" is unfortunate because physicists define an event as an unextended "point" in spacetime. It would have been better to speak of a time interval.

[68] Ibid., p. 53.

> ready by Rabbi Bachya ben Asher in his fourteenth-century commentary on Genesis, where he wrote: "For those days were not like human days, but they were the days from which one formed the unfathomable years, in a similar sense to the verse (Job 36:26) "Behold God is mighty beyond our knowledge: the number of his years is unfathomable," and it says (Job 10:5) "Are your days as the days of man?" and (Psalms 102:27) "Your years end not."

According to Domb, this interpretation is fully conformable with the relativistic conception of the nature of time as illustrated, for example, by the famous clock paradox. He concluded that "for human beings time scales are relative, and it would seem that the absolute time used at the beginning of Genesis must be related to God's time scale."[69]

Theological applications of time dilation are not confined to biblical apologetics; the relativistic effect has been used also in defense of particular theological theses. As an example, Alan G. Padgett's defense of his "new doctrine of eternity" is intimately related to the preceding discussion of the notion of eternity. In order to understand this issue, some preliminary explanatory comments are required.

The time registered by "the Eternal clock," to use Schroeder's expression, should not be confounded with what cosmologists call "cosmic time." The latter is defined at any event in space-time as the time coordinate in that reference frame, relative to which the expansion of the universe is isotropic. It can be shown that, although these reference frames corresponding to different locations are in motion

[69] C. Domb, "Faith and Reason in Judaism," in N. Mott, *Can Scientists Believe?* (James & James, London, 1991), pp. 129–142.

relative to each other, an appropriate correlation of time-defining clocks in every frame leads to a frame-independent time, the cosmic time. Assuming the existence of cosmic time, we shall see how Padgett, by appealing to time dilation, defended his thesis against objections based on the notion of cosmic time.[70]

Padgett reexamined the question of whether God's mode of existence is timeless, as Plotinus, Boethius, and more recently Stump and Kretzmann contended, or whether He exists in an everlasting duration through time. In agreement with process theology, Padgett claimed that God's relation to the world changes. But "necessarily, if a change occurs, then a duration occurs." Hence, God must in some way be temporal. The idea that God is not absolutely timeless is, however, unsatisfactory because God, as the infinite Creator of all things, also created time and must therefore in some way transcend time. One aspect of this transcendence is "relative timelessness," which means that God is timeless in the sense that He is not in "measured time," the time defined and calibrated by regular physical processes or clocks in accordance with Einstein's definition of time. This proposal has been challenged by Craig on the grounds that the cosmic time of our universe may act itself as a kind of "clock" in order to measure the duration of God's time;[71] even though God is not subject to the order of nature, cosmic time can serve to measure some duration in eternity.

To rebut this objection, Padgett assumed that God per-

[70] A. G. Padgett, "Divine Eternity and the Nature of Time," Ph.D. diss. (Oxford University, 1988); "God and Time: Toward a New Doctrine of Timeless Eternity," *Religious Studies* 25 (1989): 209–215; *God, Eternity and the Nature of Time*, chap. 6.

[71] W. L. Craig, "God and Real Time."

ceives two events e_1 and e_2 that occur at the same object and that their temporal difference in cosmic time is one "stund," where a "stund" is a measure or unit of cosmic time. Because God must exist in order for e_1 and e_2 to exist, and because e_1 and e_2 are one "stund" apart, it would follow according to the objection that God lived for one "stund" so that God cannot be timeless even in a relative sense. But retorted Padgett,

> We know that e_1 is in the light cone of e_2, since they are episodes of the same objects so that there is a causal link between them which established a conical order. But because of the well known fact of time dilation, an observer moving at a velocity near c relative to her basic frame *will not* measure the duration between e_1 and e_2 as one stund. If, then, for different observers in our own spacetime the difference between e_1 and e_2 is not always one stund, how can we insist (as the argument above does) that the duration between e_1 and e_2 will be one stund *in God's time*? Such a conclusion absolutizes our cosmos, as though God could not create a thousand such universes, all with different times. . . . Thus the argument above fails, and we can conclude that, indeed, God transcends any Measured Time, and is thus relatively timeless.[72]

Hitherto we have confined our discussion of theological implications of Einstein's scientific work to his seminal paper on relativity, which had been received by the editors of the *Annalen der Physik* on June 30, 1905. Three months later, on September 27, 1905, they received the three-page article

[72] A. G. Padgett, *God, Eternity and the Nature of Time*, pp. 128–129.

"Ist die Trägheit eines Körpers von seinem Energieinhalt abhängig?" in which Einstein derived the mass-energy relation from the electrodynamic part of his first paper.[73] This relation is usually expressed by the famous equation $E = mc^2$, where E denotes the energy of a body and m its mass. When Einstein summarized its contents by stating that "the mass of a body is a measure of its energy content," he was not aware that this statement marked the beginning of a new era in the history of civilization, the age of nuclear energy with its dangers and promises for the human race.

Man's mastery of nuclear power has been associated with theological ideas. For example, in the following statement: "According to . . . popular devotion God is conceived primarily in terms of power. The power of God is an extrapolation of earthly power: a power which, since Hiroshima, appears to be ever increasing in the hands of man."[74] In a similar vein and on the interpretation of $E = mc^2$, that mass is not only a measure of energy but ultimately energy itself, it has been claimed that "if energy is the essential basis of the whole material world, this to the Chris-

[73] A. Einstein, "Ist die Trägheit eines Körpers von seinem Energieinhalt abhängig?" *Annalen der Physik* 18 (1905): 639–641; "Does the Inertia of a Body Depend upon Its Energy Content?" in A. Einstein et al., *The Principle of Relativity* (Dover, New York, 1952), pp. 69–71. The original paper is reprinted in *The Collected Papers of Albert Einstein* vol. 2 (Princeton University Press, Princeton, N.J., 1989), pp. 312–314; and its English translation in the Princeton translations project, vol. 2, pp. 172–175. For the history and philosophical importance of this paper, see M. Jammer, *Concepts of Mass in Classical and Modern Physics* (Harvard University Press, Cambridge, Mass., 1961), chap. 13.

[74] D. Stanesby, *Science, Reason and Religion* (Routledge, London, 1988), pp. 128–129.

tian is a clear manifestation of the active creative spirit of God in the physical realm."[75]

The equation $E = mc^2$ has also been used to resolve an age-old cosmological problem of theological importance. In order to understand its proposed solution, one must refer again to some cosmological consequences of the general theory of relativity. The theological issue in question is the conflict between the traditional doctrine of God's creation of the world out of nothing, *creatio ex nihilo*, and the physical law of the conservation of energy or mass. In 1958, Henry Margenau proposed a solution to this conflict by appealing to the mass-energy relation. In a study of Thomas Aquinas in the light of modern physics, Margenau declared that Thomas's affirmation in his *De Potentia*, "that God can and does make something from nothing should be steadfastly held," was anathema to the scientists of the last century.[76] But Margenau continued, "relativity has changed all this, and it is a curious fact, perhaps not widely known, that creation of matter out of nothing contradicts no physical conservation law."[77] (Margenau used cgs units and M for mass.)

Following Margenau, consider, for the sake of simplicity, Einstein's 1917 cosmological model of a finite spherical universe of radius R, filled with an aggregation of matter

[75] A. Smethurst, *Modern Science and Christian Belief* (J. Nisbet, London, 1955).

[76] "Dicendum, quod tenendum est firmiter, quod Deus potest facere aliquid ex nihilo et facit," *S. Thomas Aequinatis Quaestiones Disputatae*, vol. 2 (Marietti, Rome, 1958), question 3, article 6, p. 14.

[77] H. Margenau, *Thomas and the Physics of 1958: A Confrontation* (Marquette University Press, Milwaukee, 1958), p. 41.

of total mass M so that its gravitational potential energy is the negative quantity $-kGM^2/R$, where G is the gravitational constant (6.67×10^{-8} cm^3 gm^{-1} sec^{-2}) and k a positive numerical factor not greatly different from 1.[78] "Because of the mass-energy equivalence, the aggregation of matter also contains the positive energy Mc^2." The total energy of the universe is therefore

$$E = Mc^2 - (kGM^2/R). \tag{12}$$

Creation from nothing is energetically possible if $E = 0$ or, equivalently, if

$$kGM/Rc^2 = 1. \tag{13}$$

Because "the following values seem to be among the best available:

$$G/c^2 = 10^{-28} \text{ cm gm}^{-1}, M = 10^{55} \text{ gm, and } R = 10^{28} \text{ cm,"} \tag{14}$$

relation (13) is indeed fulfilled for k having a value of about 10. Margenau therefore declared: "The remarkable fact is that this relation may well be fulfilled by our actual universe. . . . One truly wonders whether this is an accident!" Or, in short, Einstein's mass-energy relation fully vindicates Thomas Aquinas' dogmatic dictum.

Margenau admitted that in his simple computation he ignored the mass of dust in interstellar space, the energy of radiation and of cosmic rays, as well as interactions except gravitational attraction; but he claimed that "it is unlikely that these effects will alter the result significantly." That in

[78] A. Einstein, "Cosmological Considerations on the General Theory of Relativity."

a closed homogeneous universe the gravitational energy cancels the mass energy also can be shown without resorting to numerical values of *R* and *M*. As shown by Nathan Rosen and others in 1994, it suffices to use the pseudotensor of the gravitational field, as formulated in Einstein's 1916 fundamental paper on general relativity.[79]

One of the most remarkable attempts to connect the mass-energy relation with theological thought is its use as a scientific exegesis of Incarnation, the manifestation of the deity in a physical body. As is well known, Incarnation, the doctrine that God (or "Logos") "became flesh and dwelt among us," as stated in the Bible (John 1:14), is a basic tenet of the Christian faith. The doctrine also plays an important role in other religions; for example, in Hinduism, Vishnu, the supreme god of the Vaishnava sect, is held to have appeared in nine different incarnations or "avatars"; in Laism, the Tibetan version of Buddhism, the Dalai Lama is a personification of god. Although in principle the following considerations apply *mutatis mutandis* [with necessary changes being made], to all doctrines of Incarnation, the possibility of relating Incarnation to scientific theories has been explored only with respect to Christianity and

[79] N. Rosen, "The Energy of the Universe," *General Relativity and Gravitation* 26 (1994): 319–321. See also F. I. Cooperstock and M. Israelit, "The Energy of the Universe," *Foundations of Physics* 25 (1995): 631–635.

A. Einstein, "Die Grundlagen der allgemeinen Relativitätstheorie," *Annalen der Physik* 49 (1916): 769–822, eq. (56) on p. 809; "The foundation of the general theory of relativity," in H. A. Lorentz et al., *The Principle of Relativity*, pp. 109–164; *The Collected Papers of Albert Einstein*, vol. 6 (Princeton University Press, Princeton, N.J., 1996), pp. 284–337.

primarily by the Scottish Reformed theologian Thomas Forsyth Torrance, who was greatly influenced by Einstein. In fact, Torrance acknowledged his indebtedness to Einstein for having "taught [him] much . . . about coordinating reflection with realities of experience."[80]

Torrance is the champion of what he calls "theological science," which tries not only to reconcile, but also to integrate and unify, religion and science with respect to both method and contents. But he warns us that theological science should not be confounded with "natural theology" which to endorse "would be to open the floodgates of naturalism and inundate the Church with paganism."[81] Among his numerous publications, the one that deals specifically with our present topic is the book *Space, Time and Incarnation*.[82] Because of Einstein, he wrote, "For the first time . . . in the history of thought, Christian theology finds itself in the throes of a new scientific culture that is not antithetical to it, but which operates with a non-dualistic outlook upon the universe which is not inconsistent with Christian faith, even at the crucial points of creation and incarnation." According to Torrance, the notion of Incarnation is intimately related to the concepts of space and time. Torrance argued that the classical Newtonian concept of absolute space, which exists even in the absence of matter, gave rise to a mechanistic worldview that was more detrimental to science than to religion. In contrast, as conceived

[80] T. F. Torrance, *God and Rationality* (Oxford University Press, London, 1971), Preface, p. viii.

[81] T. F. Torrance, "The Word of God and the Nature of Man," in F. W. Camfield, ed., *Reformation Old and New* (Butterworth, London, 1947), p. 136.

[82] T. F. Torrance, *Space, Time and Incarnation* (Oxford University Press, London, 1969).

in Einstein's general theory of relativity, space and time are dynamic because they depend, via the field equations, on the changing distributions of mass and energy and are therefore accessible to a constructive dialogue with religion. Classical physics is based on a concept of space as a receptacle or vessel that can be filled with something from the "outside," whereas the Einsteinian concept of space (or space-time) is that of a field of relations.[83]

Torrance was certainly right when he wrote that Einstein's

> conception of space and time is the relational idea which was given its supreme expression in the space-time of relativity theory, when Einstein, following out a line of thought from four-dimensional geometry, found he had to reject the notion of absolute space and time, both as taught by Kant, for whom they were a priori forms of intuition outside the range of experience, and as taught by Newton, for whom they formed an inertial system independent of material events in them, but acting on them and conditioning

[83] Cf. in this context Einstein's Preface to M. Jammer, *Concepts of Space: The History of Theories of Space in Physics* (Harvard University Press, Cambridge, Mass., 1954, 1969; enlarged edition, Dover, New York, 1993), which contains the following statement, the last he made in his life, about the nature of space: "The victory over the concept of absolute space or over that of the inertial system became possible only because the concept of the material object was gradually replaced as the fundamental concept of physics by that of the field. . . . If the laws of this field are in general covariant, that is, are not dependent on a particular choice of coordinate system, then the introduction of an independent (absolute) space is no longer necessary. That which constitutes the spatial character of reality is then simply the four-dimensionality of the field. There is then no 'empty' space, that is, there is no space without a field." Pp. xv–xvi.

> our knowledge of the universe. This had the effect of shattering the receptacle idea and of undermining the radical dualism to which it had given rise in modern philosophy and theology as well as science.[84]

This statement by Torrance has been interpreted by Frank D. Schubert in the spirit of scientific theology as follows:

> This shattering has resulted . . . in the establishment of a new paradigm of field theory in which the traditional barrier between God and man (the "wall" of the vessel) is replaced by an essential "relatedness." It is clear that Einstein's theory provides this very relatedness . . . for it seems to point to an equation which can, at the same time, account for energy and matter as true functions of each other ($E = mc^2$). A God of pure energy, then, could well become "matter" in an Incarnation.[85]

This scientific explication of Incarnation, stating that energy "could well *become* 'matter,'" obviously interprets the mass-energy relation in the sense that energy and mass are interconvertible, that is, that energy is annihilated and mass created, or vice versa, and therefore must cope with the conceptual difficulty that energy and mass have different physical dimensions. Another interpretation of the mass-energy relation that avoids this difficulty and is shared by the majority of physicists as well as by Einstein, regards the equation $E = mc^2$ as the expression of merely a proportionality

[84] T. F. Torrance, *Space, Time and Incarnation*, p. 58.

[85] F. D. Schubert, "Thomas F. Torrance: The Case for a Theological Science," *Encounter* 45 (1984): 123–137, quotation on p. 133.

between two attributes, *E* and *m*, of one and the same ontological substratum without the occurrence of any annihilation or creative process.[86] Furthermore, incarnation, as conceived in Christian theology, implies some localization in space and time ("became flesh and dwelt among us"), an issue that is completely ignored in the proposed scientific explanation.

It also should be mentioned that Torrance's attempt to construct a scientific theology on the basis of Einstein's general theory of relativity—in particular, on his conception of space and time as a field, serving the role of a medium for the interaction of God with the physical world and humanity—has been criticized.[87] Wing-Hong Wong said that Torrance's effort was only partly successful, and the physicist and Anglican priest John Polkinghorne said that Torrance's work was "dangerously close to Newton's equation of absolute space with the sensorium of God."[88] In addition, the noted theologian Wolfhart Pannenberg pointed out that "the relation of space and time to the dynamic of the Spirit needs closer definition."[89] Pannenberg

[86] For interpretations of the mass-energy relation, see M. Jammer, *Concepts of Mass in Contemporary Physics and Philosophy* (Princeton University Press, Princeton, N.J., forthcoming), chap. 3.

[87] Torrance has spent much time in studying the conceptual development of field theories from Faraday through Maxwell to Einstein. He even published a new edition with explanatory comments of J. C. Maxwell's famous essay, "A Dynamical Theory of the Electrodynamic Field" (Scottish Academic Press, Edinburgh, 1982).

[88] W.-H. Wong, "An Appraisal of the Interpretation of Einsteinian Physics in T. F. Torrance's Scientific Theology," Ph.D. diss., University of Aberdeen, 1992.

J. Polkinghorne, *Reason and Reality: The Relationship between Science and Theology* (Trinity Press International, Philadelphia, 1991), p. 93.

[89] W. Pannenberg, *Systematische Theologie* (Vandenbroeck & Rup-

did not dispute Torrance's contention that the field concept has theological connotations because of God's continuing sustenance of the created world in space and time. On the contrary, according to Pannenberg,

> the turn toward the field concept in the development of modern physics has theological significance. This is suggested not only by its opposition to the tendency to reduce the concept of force to bodies or masses but also because field theories from Faraday to Albert Einstein claim priority for the whole over the parts. This is of theological significance because God has to be conceived as the unifying ground of the whole universe if God is to be conceived as creator and redeemer of the world. The field concept could be used in theology to make the effective presence of God in every single phenomenon intelligible."[90]

According to Pannenberg, the Biblical use of the term "spirit" justifies the assignment of theological significance to Einstein's concept of field. The Old Testament says that "the spirit of God was moving over the face of the waters"[91] and the New Testament is even more explicit in saying that "God is spirit" ("God is *pneuma*").[92]

To prove his assertion, Pannenberg referred to an article on the history and philosophy of the notions of "field" and "field theory," in which it is shown that the idea of an om-

---

recht, Goettingen, 1991); *Systematic Theology*, vol. 2 (T. & T. Clark, Edinburgh, 1994), chap. 2.

[90] W. Pannenberg, "The Doctrine of Creation and Modern Science," *Zygon* 23 (1988): 3–21.

[91] "W'ruah elohim m'rahefet al-pnei hamajim," *Genesis*, 1:2.

[92] "Pneuma ho theos," *The Gospel according to John*, 4:24.

nipresent and all-permeating *pneuma* of the Stoics (Zenon, Chrysippos, Poseisonios) anticipated the modern concept of field.[93] Pannenberg concluded therefore that "the Stoic doctrine of the divine pneuma was actually the direct precursor of the modern field concept."[94]

The justification for ascribing a theological significance to the concept of field, as proposed by Torrance and Pannenberg, has been challenged by the biochemist Jeffrey S. Wicken as "being overly bound to physical science" and being based on a misinterpretation of physical theory. "Before Albert Einstein," Wicken argued, "the field of space was hypostatized to an ether filling the void. After Einstein, the void ceased to exist, and space itself *became* a field, the central connective tissue by which forces were transmitted." According to Pannenberg, Wicken continued,

> this dematerialization of the field gives God (being immaterial) a kind of physical justification in nature's wholeness. Although a metaphor, this notion is rich for theology; taken literally, it binds God needlessly to physics. Is God conceived here as a *field as in physics*? If so, why the need for God at all? If not, the relativistic reification of the space-field seems important to theology only in the sense of showing that nature has its own grounds for wholeness that might provide boundary conditions for God's presence in nature.[95]

[93] M. Jammer, "Feld—Feldtheorie," in J. Ritter, ed., *Historisches Wörterbuch der Philosophie* vol. 2 (Schwabe, Basel, 1972), pp. 923–927; "Feld," *Archiv der Geschichte der Naturwissenschaften* (Hollinek, Wien, 1981), pp. 113–121.

[94] W. Pannenberg, *Systematic Theology*, p. 81.

[95] J. S. Wicken, "Theology and Science in the Evolving Cosmos: A Need for Dialogue," *Zygon* 23 (1988): 45–55.

Wicken criticized in particular Pannenberg's assignment of ontological priority to the field as a "whole" exerting regulative influence (God's hand) over matter. Pannenberg ignored the idea that field and matter constitute the "whole," the structure of the field being regulated (according to the field equations) by matter (or mass-energy, the energy-momentum tensor). "If all the matter were removed from the universe, there would be no field. . . . Space and matter have *coevolved*, and are *relationally constituted* by each other. They have no identities apart from each other."[96] Summing up his critique, Wicken wrote, "If we want to use the word 'field' in science-theology discourse, let us do so in some way commensurate with their understandings in physics. Talking about 'spirit' as 'energy' and granting it by implication the status of physical law runs dangerously close to usurping the hard-won denotative language of science for physicalizing theology. This serves neither enterprise."[97]

THE TORRANCE-PANNENBERG thesis that modern field theory has theological relevance, in contrast to the antireligious implications of the classical physics of mass particles moving through absolute space, was also challenged by Hans-Dieter Mutschler, as being based on merely a metaphor.[98] According to Mutschler, the association of physical fields with spiritual or incorporeal beings has its origin in an evolutionary accident. Some fish are sensitive to electri-

[96] Ibid., p. 52.

[97] Ibid., p. 48.

[98] H.-D. Mutschler, "Schöpfungstheologie und physikalischer Feldbegriff bei Wolfhart Pannenberg," *Theologie und Philosophie* 70 (1995): pp. 543–558.

cal fields, and some birds are sensitive to magnetic fields. Evolution could well have proceeded so that the human species would also have developed organs capable of sensing physical fields and of perceiving them as clouds, mists, or sandstorms. We would then never associate them with spiritual entities. The relation "spirit to matter like field to particle" [Geist zu Materie wie Feld zu Partikel] is merely an analogy and not an identification of the corresponding concepts involved. Furthermore, the Poisson equation of classical physics or the field equations of the general theory of relativity, which correlate ponderable matter with fields, are "readable in both directions" and do not assign ontological priority to the notion of field, just as the mass-energy relation $E = mc^2$ does not ascribe ontological priority to energy over mass.

In a rejoinder to Mutschler's critique, Pannenberg admitted that his contention cannot be derived from physical science in a purely logical way. Nevertheless, he continued, it has a philosophical foundation stronger than that presented by Mutschler because it takes into consideration the notions of space and time. Apart from this, it would be wrong to regard the language of physics as being totally independent of philosophical interpretations as Mutschler believes. Even if such independence may hold for the mathematical formulae, the fundamental concepts applied in them always remain entwined with their historical origins. The classical tendency to reduce operations of forces to material bodies left no room for the conceptualization of a creative interaction between God and nature because of the traditional and well-founded conviction that God, whatever His nature may be, is certainly not a body. The introduction of the field concept into physics therefore opens the possibility of a new conception of the relation between physics and theology.[99]

Mutschler was certainly right in warning us not to confound a fundamental notion of a scientific theory, like the notion of a field in Einstein's relativity, with a metaphor or anticipation of it in, or a projection of it into, earlier systems of thought. If we accept, however, the view that there persist throughout the history of scientific thought certain ideas, patterns, or paradigms that may have been influential, even if only subconsciously, on the construction of a new theory, then a study of such anticipations can provide some information about the ideological background that supported the formation of the new theory. In the context of the present study, it seems, therefore, appropriate to discuss briefly those anticipations of fundamental concepts of relativity that have some relevance to religion or theology.

According to a historical treatise on ancient Chinese medicine published by Georges Beau in 1965, a wise man at the court of a Chinese emperor anticipated not less than 4,500 years ago Einstein's notion of space-time and his mass-energy relation.[100] But as Shigeru Nakayama and Nathan Sivin have shown, Beau's contention is "based on a misinterpretation of a ludicrously misdated text."[101]

However, a leading expert on the history of Chinese civilization, Joseph Needham, studied ancient concepts of space and time and found that *yü-chou*, frequently used in

[99] W. Pannenberg, "Geist als Feld—nur eine Metapher?" *Theologie und Philosophie* 71 (1996): 257–260.

[100] G. Beau, *La Médecine Chinoise* (Éditions du Seuil, Paris, 1965), p. 182.

[101] S. Nakayam and N. Sivin, eds., *Chinese Science* (MIT Press, Cambridge, Mass., 1973), p. xv.

a second-century Chinese compilation of ancient philosophy, has the meaning of Einstein's (or Minkowski's) space-time. Needham quoted a text from 120 B.C., entitled *Huai Nan Tzu*: "All the time that has passed from antiquity until now is called *chou*; all the space in every direction, above and below, is called *yü*. The Tao (the Order of Nature) is within them, yet no man can say where it dwells." And Needham explained, "The original meaning of both these two ancient words was 'roof,' of house, cart or boat, so that the semantic significance is that of something stretching over an expanse to cover it. So indeed we still in English say that such and such an exposition 'covers' ten or fifteen centuries."[102] A similar interpretation has been given to the biblical term *'olam*, which was originally a purely temporal notion denoting "a time in the distant past," but later assumed the meaning of both space and time and finally the meaning of the Greek notion *cosmos*, as used, for example, in God's appellation as *meleḥ ha'olam* ("King of the world").[103]

TURNING NOW to anticipatory theories of the concept of field in relativity, we note that the above-mentioned proposed derivation of this notion from the idea of a divine *pneuma* can be regarded as such a theory. A similar suggestion has been made by the theoretical physicist Mendel Sachs who has thoroughly studied the history of the concept of field and written a monograph on this subject. Sachs, an enthusiastic aficionado of the informative importance of anticipa-

[102] J. Needham, "Time and Knowledge in China and the West," in J. T. Fraser, ed., *The Voices of Time* (George Braziller, New York, 1966), pp. 92–135, quotations on p. 93.

[103] E. Jenni, "Das Wort 'olam im Alten Testament," *Zeitschrift für die alttestamentliche Wissenschaft* 64 (1952): 197–248; 65 (1953): 1–25.

tions, or of what he calls "invariant ideas with respect to change from one contextual framework to another," claimed that the relativistic field concept has much in common with Moses Maimonides' idea about resolving the problem of how God, in spite of His incorporality, can exert His influence on corporal entities of an inanimate and living nature.[104] In "On the Nature of Divine Influence," chapter 12 of *The Guide for the Perplexed*, Maimonides wrote:

> There are actions that do not depend on impact, or on a certain distance. They are termed "influence" (or "emanation"), on account of their similarity to a water-spring. The latter sends forth water in all directions, has no peculiar side for receiving or spending its contents; it springs forth on all sides, and continually waters both neighbouring and distant places. In a similar manner, incorporeal beings, in receiving power and imparting it to others, are not limited to a particular side, distance, or time. They act continually; and whenever an object is sufficiently prepared, it receives the effect of that continuous action, called "influence" (or "emanation"). God being incorporeal, and everything being the work of Him as the efficient cause, we say that the Universe has been created by the Divine influence, and that all changes in the Universe emanate from Him.[105]

According to Sachs, these statements of Maimonides anticipated Einstein's field concept and "the correspondence of

[104] M. Sachs, *The Field Concept of Contemporary Science* (Charles C. Thomas, Springfield, Ill., 1973).

[105] M. Maimonides, *The Guide for the Perplexed*, trans. M. Friedländer (George Routledge & Sons, London; E. P. Dutton, New York, 1904), p. 170.

Maimonides' ideas . . . with those of Einstein . . . strengthen the view that indeed there are fundamental ideas about the real world that persist throughout the history of mankind."[106] Sachs also compared the relativistic concept of field with certain philosophical notions of Far-Eastern religions and, in particular, with the Buddhistic conception of "anatmān," the denial of the "self" or the "ego."[107]

The earliest Christian writer, whose concepts have been claimed to accord with the modern notion of space and time, is probably the ninth-century theologian John Scotus Eriugena (Erigena). Certain statements in his highly original work *De Divisione Naturae (Periphyseon)*, like "omne temporale locale sit et conversim omne locale temporale" or "omnium . . . existentium essentia localis atque temporalis est"[108] have been interpreted as an assertion of an intrinsic relationship between space and time.[109]

Quite a few exponents of neo-Thomism declare that Thomas Aquinas expressed in his writings, particularly his *Summa Theologiae*, which still forms the basis of the dog-

[106] M. Sachs, "Maimonides, Spinoza, and the Field Concept in Physics," *Journal of the History of Ideas* 37 (1976): 125–131.

[107] M. Sachs, "On the Philosophy of General Relativity Theory and Ideas of Eastern and Western Cultures," in S. Fujita, ed., *Festschrift for Ta-You-Wou* (Gordon and Breach, New York, 1978), pp. 9–24.

M. Sachs, "Comparison of the Field Concept of Matter in Relativity Physics, and the Buddhist Idea of Nonself," *Philosophy East and West* 33 (1983): 395–399.

[108] *Joannis Scoti Erigenae De Devisione Naturae Libri Quinque* (Theatro Shelsoniano, Oxford, 1681; Minerva, Frankfurt, 1964), book 5, chap. 17.

[109] "Erigena ist mit diesen Ausführungen der erste, der einen *inneren* Zusammenhang zwischen Raum und Zeit aufweist. Beide bilden *gemeinsam* die Bedingungen für die ontologische Seinsmöglichkeit der Erscheinungswelt." A. Gosztony, *Der Raum* (K. Albers Verlag, Freiburg, 1976), p. 149.

matic teachings of the Catholic Church, certain ideas that, at least from the methodological point of view, are quite similar to some concepts in Einstein's theory of general relativity. As is well-known, Einstein's theory is also a theory of gravitation. Such a claim has been made by James A. Weisheipl in his article "Space and Gravitation."[110] Weisheipl stated that if we distinguish between mathematical theory and natural philosophy, then the theory of nature and gravitation as taught by Aquinas "offers a realistic basis and justification for the theory of relativity in its essential content." William A. Wallace contended in a similar vein that "the essential contribution of Einstein is to cancel out the excessive mathematical realism of Galileo, while still leaving open the possibility of a type of physical certainty and proof as conceived by Thomas Aquinas."[111] These and other parallelisms have been further elaborated in John F. Kiley's book *Einstein and Aquinas: A Rapprochement.*[112] When G. S. Viereck told Einstein, in the course of his above-mentioned interview, that a Dublin periodical had published an article by a Catholic theologian stating that the theory of relativity "merely confirms the teachings of Thomas Aquinas," Einstein replied: "I have not read all the works of Thomas Aquinas, but I am delighted if I have reached the same conclusions as the comprehensive mind of that great Catholic scholar."[113]

Most remarkably in 1601, Johann Kepler anticipated one

[110] J. A. Weisheipl, "Space and Gravitation," *The New Scholasticism* 29 (1955): 175–223.

[111] W. A. Wallace, "St. Thomas, Galileo, and Einstein," *Thomist* 24 (1961): 1–22.

[112] J. F. Kiley, *Einstein and Aquinas; A Rapprochment* (M. Nijhoff, The Hague, 1969).

[113] G. S. Viereck, *Glimpses of the Great*, p. 374.

of the the central ideas of Einstein's general theory of relativity. At that time, alchemistic and astrological ideas, inspired by religious considerations, played an important role in astronomical investigations. In this context, Kepler, in a study of the foundations of astrology, made the statement "ubi materia, ibi geometria" ("where matter is, there is geometry").[114] This statement may be interpreted as an ingenious epigram of the physical contents of the field equations of general relativity, which relate the geometrical structure of space-time to the mass-energy (or momentum-stress) density of the universe.

WITHOUT INVOLVING "time," some theologians have argued that the notion of four-dimensionality is as old as the Bible because St. Paul's Epistle to the Ephesians referred to "breadth and length and height and depth."[115] The problem of why space has only three dimensions—if certain modern higher-dimensional theories are ignored—and why time has only one dimension is, of course, a topic widely discussed among philosophers and physicists, but need not be dealt with in our present context.[116] Let us only mention in passing the little known fact that Immanuel Kant's 1770 dissertation anticipated Hermann Minkowski's geometrical representation of relativistic space-time. "If time is represented," wrote Kant, "by an infinitely long straight line and all simultaneous occur-

[114] J. Kepler, "De fundamentis astrologiae certioribus," thesis XX in *Johann Kepler, Gesammelte Werke*, vol. 4 (Beck, München, 1941), p. 15.

[115] "To platos kai mēkos kai hypsos kai bathos," *Ephesians*, 3:18. See also *Job*, 11:7–9.

[116] See, e.g., M. Jammer, "Science and Philosophy in the Problem of the Dimensionality of Space," in Y. Elkana, *The Interaction between Science and Philosophy* (Humanities Press, Atlantic Highlands, N.J., 1975), pp. 463–468.

rences at a given moment are represented by a transversely drawn straight line through that point on the time line, the thus generated surface represents the phenomenal world with respect both to its substance and its accidents."[117] As a Newtonian, Kant of course did not yet envisage the possibility of mutually inclined "infinitely long straight lines" representing the (proper) time along different world lines of different observers in relative motion to each other that characterizes relativistic space-time.

I shall not discuss the role of the fourth dimension in the spiritualistic and theosophical literature associated with the names of F. Oetinger, J. Fricker, W. Crookes, K. F. Zöllner, H. P. Blavatsky, C. H. Hinton, P. D. Ouspensky, and others, before and after the advent of Einstein's theory of relativity. Sometimes the scientific terminology more than the physical contents of relativity prompted people to draw theological conclusions from Einstein's theory. Thus, for example, in his 1939 article on "Immortality and a Fourth Dimension," Henry H. Riggs, Director of the Department of Systematic Theology at the Near Eastern School in Beirut, Syria, wrote: "The mathematical concept from which I think that we may find some light on the reality of eternal life is familiar to mathematicians as the Fourth Dimension; and the rather nebulous idea has been brought to the front and set to work in connection with the theory of Relativity, in the proposition, built into the very foundation of that

[117] I. Kant, *De mundi sensibilis atque intelligibilis forma et principiis*, vol. 2 (Akademie Ausgabe, G. Reimer, Berlin, 1912), pp. 385–419; W. J. Eckoff, ed., *Kant's Inaugural Dissertation of 1770* (Columbia College, New York, 1894); J. Handyside, *Kant's Inaugural Dissertation and Early Writings on Space* (The Open Court, Chicago, 1929).

theory, that time is the fourth dimension."[118] On the basis of a similar argument, Sir Richard Tute, a chief justice of the Bahamas, arrived at the conclusion that

> science has done religion, all the higher religions, a great service, by helping to establish the existence of a state, in the shape of space-time, in which survival can take place. It has probably done more, since it would appear to have established the existence of a state in which survival must take place. . . . All this is wholly in accord with the religious outlook. It is curious to reflect that scientists, who had no other motive than the elucidation of a puzzle of nature, should have stumbled on a truth which is essential to the interpretation of religion as it is to the interpretation of the physical world.[119]

Einstein would certainly not have agreed with such interpretations of the fourth dimension, not only because of their obscure logic, but simply because he did not believe in life after death. In fact, in 1953, Einstein received a letter from Alice M. Nickerson, a licensed Baptist pastor in Manchester, New Hampshire, which contained several quotations from the New Testament (Mark 8: 30–37; Romans 3: 23; II Peter 3: 9) and the question of whether he had ever considered "the relation of [his] immortal soul and its salvation to the Creator who gave it." We do not know what Einstein replied, if he did so at all. The sender's letter, pre-

[118] H. H. Riggs, "Immortality and a Fourth Dimension," *The Hibbert Journal* 37 (1938/1939): 264–270.

[119] R. Tute, "Space Time—A Link between Religion and Science," *Hibbert Journal* 38 (1939/1940): 261–270.

served in the Einstein Archive, carries in Einstein's handwriting, and in English, the statement, "I do not believe in immortality of the individual."[120]

HITHERTO WE have discussed only references to the special and the general theory of relativity supporting the claim that Einstein's scientific work has theological significance. But Einstein's contributions to other branches of physics were also of decisive importance. As every student of physics knows, Einstein was one of the founding fathers of quantum theory. In fact, his early work on light quanta and the photoelectric effect earned him the Nobel Prize in 1921. His later critical work on quantum mechanics greatly affected the progress of this theory.[121] "Were it not for Einstein's challenge," declared Niels Bohr in 1961, "the development of quantum physics would have been much slower." His work on light quanta, on quantum statistics, on the specific heat of solids, and his "astonishingly simple" 1916 derivation of Planck's radiation law, in which he introduced his famous probability coefficients for spontaneous and induced transitions, have never been claimed to have any theological im-

[120] A. M. Nickerson to Einstein, 17 July 1953. Einstein Archive, reel 36-552. This statement is corroborated by the fact that, when once asked whether he had faith "in a life to come," he replied: "No, I have faith in the universe, for it is rational. And I have faith in my purpose here on earth. I have faith in my intuition, the language of my conscience, but I have no faith in speculation about Heaven and Hell." W. Hermanns, *Einstein and the Poet* (Branden Press, Brookline Village, Mass., 1983), p. 94.

[121] For a more detailed survey of Einstein's contributions to the quantum theory, see A. Pais, *Subtle is the Lord* (Oxford University Press, Oxford, 1982), chap. 6; M. A. El'yashevich, "Einstein's Part in the Development of Quantum Concepts, *Uspekhi Fisiceskich Nauk* 128 (1979): 503–536; *Soviet Physics Uspekhi* 22 (1979): 555–575.

port. Therefore, I examine the more philosophical aspects of his later critique of quantum mechanics, the impact of which still affects present-day research. First, I briefly review Einstein's critique of modern quantum mechanics and its effect on recent developments of this theory.

Quantum mechanics is based on Heisenberg's indeterminacy relations, according to which canonically conjugate observables—that is, physical quantities like position q and momentum p, or energy E and time t—cannot simultaneously be determined sharply. It is also based on Born's interpretation of the wave function as a measure of the probability of finding the particle in the state characterized by that function. This probability is conceived as irreducible and inherent in nature and not as an expression of human ignorance of the details of elementary processes as in classical statistical mechanics. Born repeatedly admitted that his probability interpretation had been influenced by Einstein's conception of the relation between light quanta and the electromagnetic field, whose squared wave amplitudes determine the probability of the presence of photons.[122]

Because Einstein was deeply influenced by Spinoza and his philosophy of determinism, it is not surprising that Einstein never accepted the quantum mechanical abandonment of determinism and causality. As early as 1912, he wrote to his friend Heinrich Zangger, "The more success the quantum theory has, the sillier it looks."[123] In 1926, he wrote to

[122] See M. Jammer, *The Philosophy of Quantum Mechanics—The Interpretations of Quantum Mechanics in Historical Perspective* (Wiley, New York, 1974), pp. 40–41.

[123] "Je mehr Erfolg die Quantumtheorie hat, desto dümmer sieht sie aus." Einstein to H. Zangger, 20 May 1912. Einstein Archive, reel 39-655. *Collected Papers of Albert Einstein*, vol. 5 (1993), p. 467.

Born the often-quoted words, "Quantum mechanics is very worthy of regard; but an inner voice tells me that it is not the true Jacob. The theory yields much, but it hardly brings us close to the secrets of the Old One. In any case, I am convinced He does not play dice."[124] When Leopold Infeld once asked Einstein, "Why are you so dissatisfied with quantum theory, especially with the development that really started from your work?" Einstein replied, "Yes, I may have started it but I always regarded these ideas as temporary. I never thought that others would take them so much more seriously than I did."[125] What Infeld had in mind was Einstein's work on light quanta, especially his use of the probability coefficients. Einstein considered his introduction of these coefficients merely a provisional device, as shown by his concluding remark to their introduction. "The weakness of the theory lies . . . in the fact . . . that it leaves the duration and direction of the elementary processes to 'chance'."[126]

In order to resolve the paradox of the experimentally observed wave-particle duality and to explain the indeterminacy relations, Bohr propounded in 1927 his complementarity interpretation. It said, "Nature forces us to adopt

[124] "Die Quantenmechanik ist sehr achtung-gebietend. Aber eine innere Stimme sagt mir, dass das doch nicht der wahre Jakob ist. Die Theorie liefert viel, aber dem Geheimnis des Alten bringt sie uns kaum näher. Jedenfalls bin ich überzeugt, dass der nicht würfelt." Einstein to M. Born, 4 December 1926. Einstein Archive, reel 8-180.

[125] L. Infeld, *Albert Einstein—His Work and Its Influence on Our World* (C. Scribner's Sons, New York, 1950), p. 110.

[126] "Die Schwäche der Theorie liegt darin, dass sie . . . Zeit und Richtung der Elementarprozesse dem 'Zufall' überlässt." A. Einstein, "Zur Quantentheorie der Strahlung," *Mitteilungen der Physikalischen Gesellschaft in Zürich* 16 (1916): 47–62; *Physikalische Zeitschrift* 18 (1917): 121–128; *Collected Papers of Albert Einstein*, vol. 6 (1996), pp. 382–387, quotation on p. 396.

a new mode of description designated as complementary in the sense that any application of classical concepts [like "position"] precludes the simultaneous use of other classical concepts [like "momentum"] which in different connection are equally necessary for the elucidation of phenomena."[127] When asked whether quantum mechanics could be considered to mirror an underlying quantum reality, Bohr replied, "There is no quantum world. There is only an abstract quantum physical description. It is wrong to think that the task of physics is to find out how nature *is*. Physics concerns what we can *say* about nature."[128]

If we recall that Einstein, as a firmly convinced objective realist, regarded "physics as an attempt to grasp reality as it is thought, independently of its being observed," we understand that he strongly disagreed with Bohr and his "Copenhagen interpretation," which gained the acceptance of the majority of physicists.[129] It is not surprising that, in a famous debate with Bohr, Einstein presented a number of thought experiments designed to disprove the indeterminacy relations. Bohr succeeded, however, in refuting each of Einstein's challenges by consistently applying the indeterminacy relations. The peak of the debate occurred when Einstein presented, on October 22, 1930, at the Sixth Solvay Congress in Brussels, his photon-box experiment designed to disprove the energy-time indeterminacy relation. This time, after a sleepless night, Bohr rebutted Ein-

[127] N. Bohr, *Atomic Theory and the Description of Nature* (Cambridge University Press, Cambridge, 1934), p. 10.

[128] A. Petersen, "The Philosophy of Niels Bohr," *Bulletin of the Atomic Physicist* 19 (1963): 8–14, quotation on p. 12. Emphasis in original.

[129] A. Einstein, "Autobiographical Notes," p. 81.

stein's charge by using Einstein's own general theory of relativity, according to which the rate of a clock depends on the gravitational potential.[130]

Defeated, but not convinced, Einstein continued to challenge orthodox quantum mechanics by contending that, even if it is consistent, it is not complete. Because of its importance for the sequel, I briefly outline the incompleteness argument presented in the famous "EPR paper," published by Einstein in 1935 in collaboration with Boris Podolsky and Nathan Rosen. An element of physical reality is defined in it by the following sufficient condition (1): "If, without in any way disturbing a system, we can predict with certainty the value of a physical quantity, then there exists an element of physical reality corresponding to this physical quantity." The completeness of a physical theory is defined by the necessary condition (2): "Every element of physical reality must have a counterpart in the physical theory." In order to prove that (2) is not satisfied, the authors considered two particles or systems S_1 and S_2, whose position and momentum variables are denoted by q_1, q_2, and p_1, p_2, respectively. It is assumed that, after having interacted, the two systems separate from each other in a state in which their relative distance $Q = q_1 - q_2$ and their total momentum $P = p_1 + p_2$ have precise values, which is possible because Q and P commute. If then, q_1 is measured, q_2 can be predicted with certainty without disturbing S_2, so

[130] For details of the Bohr-Einstein debate, see N. Bohr, "Discussion with Einstein, in *Albert Einstein: Philosopher-Scientist*, pp. 199–241. See also M. Jammer, *The Philosophy of Quantum Mechanics* (Wiley, New York, 1974), chap. 4–6; and for a nonmathematical review, A. Whitaker, *Einstein, Bohr and the Quantum Dilemma* (Cambridge University Press, Cambridge, 1996).

that according to (1), q_2 is an element of physical reality. If, however, p_1 is measured, p_2 can be predicted without disturbing S_2, and, again according to (1), p_2 is an element of physical reality. Because q_2 and p_2 are noncommuting conjugate variables, the completeness condition (2) is not satisfied for at least one of these variables.[131]

The EPR argument provoked a host of discussions and heated debates. Reacting to them, Einstein declared, "But on one supposition we should, in my opinion, absolutely hold fast: the real factual situation of the system S_2 is independent of what is done with the system S_1, which is spatially separated from the former." From the dilemma raised by the incompleteness argument, Einstein continued, "one can escape . . . only by either assuming that the measurement of S_1 (telepathically) changes the real situation of S_2 or by denying independent real situations as such to things which are spatially separated [that is, by a spacelike separation] from each other. Both alternatives appear to me entirely unacceptable."[132]

Einstein's "supposition" that the properties of one system cannot be affected by what is done to another system in spacelike separation from the former is usually referred

[131] A. Einstein, B. Podolsky, and N. Rosen, "Can Quantum-Mechanical Description of Physical Reality Be Considered Complete?" *Physical Review* 47 (1935): 777–780. Although the central idea of this paper is undoubtedly Einstein's, its presentation in the text, written by Podolsky, did not precisely reflect Einstein's view of completeness. See A. Fine, "Einstein's Critique of Quantum Theory: The Roots and Significance of EPR," in P. Barker and C. G. Shugart, *After Einstein* (Memphis State University Press, Memphis, 1981), pp. 147–158. Because it is the published version that affected subsequent developments, we ignore these details.

[132] A. Einstein, "Autobiographical Notes," p. 85.

to as the *locality principle* or *separability principle*, or briefly as *Einstein locality*."[133] The combination of this principle with the philosophical doctrine of realism is generally called *local realism*."[134] No less than determinism, local realism is a fundamental tenet of Einstein's philosophy of science.

In 1964, John Stewart Bell made what has been hailed by Henry P. Stapp as "the most profound discovery of science," when he proved rigorously that local realism leads inescapably to certain mathematical consequences, the so-called Bell inequalities, which are violated by the predictions of quantum mechanics.[135] These violations have been convincingly confirmed in a number of experiments, most notably in those performed by Alain Aspect and his team at the Institut d'Optique Théorique et Appliquée in Paris.[136] Thus, the philosophical problem of whether local realism is maintainable has been settled in the laboratory, but alas with a negative result. This result, reached by what Abner Shimony has fittingly called "experimental metaphysics," invalidates—at least as far as present-day evidence shows —a basic tenet of Einstein's philosophy of science. The untenability of Einstein's local realism is the outcome of a

[133] For precise definitions of these notions, see D. Howard, "Einstein on Locality and Separability," *Studies in History and Philosophy of Science* 16 (1985): 171–201.

[134] For technical reasons, the definition of local realism often also contains the denial of temporal retroaction, that is, the denial that the past can be influenced by choices made in the future.

[135] J. S. Bell, "On the Einstein-Podolsky-Rosen Paradox," *Physics* 1 (1964): 195–200.

[136] A. Aspect, J. Dalibard, and G. Roger, "Experimental Test of Bell's Inequality Using Time-Varying Analyzers," *Physical Review Letters* 49 (1982): 1804–1807.

conceptual development that, ironically, owes its initiation to Einstein himself.

Not only orthodox quantum mechanics, but also most of the other proposed alternatives and even David Bohm's realistic hidden-variable theory, are nonlocal theories, the latter because of its use of the quantum potential.[137] Recall in this context that Einstein, who had never accepted the idea of hidden variables and had called them "too cheap," had nevertheless played an important role in Bohm's inception of his hidden-variables theory. In 1951, when teaching at Princeton University, Bohm wrote his well-known textbook *Quantum Theory* in full conformance with Bohr's complementarity interpretation.[138] As he later admitted, he wrote it in order to explain to himself "the precise nature of the new quantum-theoretical concepts." As soon as he had completed the book, he felt greatly dissatisfied with its approach, because it "could not go beyond the phenomena or appearances." Einstein, to whom he had sent a copy of the book, invited him for a discussion on it. "This encounter," Bohm reminisced, "had a strong effect on the direction of my research, because I then became seriously interested in whether a deterministic extension of the quantum theory could be found."[139]

Einstein's influence on Bohm's reasoning has some relevance to our topic, even though only an indirect one. The direction of Bohm's research led him from his hidden vari-

[137] D. Bohm, "A Suggested Interpretation of the Quantum Theory in Terms of 'Hidden Variables,'" *Physical Review* 85 (1952): 166–193.

[138] D. Bohm, *Quantum Theory* (Prentice-Hall, Englewood Cliffs, N.J., 1951).

[139] D. Bohm, "Hidden Variables and the Implicate Order," *Zygon* 20 (1985): 111–124, quotation on pp. 113–114.

ables to his philosophy of "implicate order," the theological implications of which were the subject of the 1983 Berkeley conference on "David Bohm's Implicate Order: Physics and Theology," sponsored by the Graduate Theological Union, and of other conferences. In 1983 Bohm delivered at St. James' Church, Piccadilly, London, a talk entitled "Fragmentation and Wholeness in Religion and in Science," in which he elaborated on the consequences of his physical theory for theology.[140]

BEFORE DISCUSSING in detail arguments that claim that Einstein's critique and his critical contributions to quantum physics are of theological importance, I wish to point out that, in contrast to the theories of relativity, quite a few of its originators considered quantum theory theologically significant. For example, in his *Physics and Philosophy*, Heisenberg emphasized that "atomic physics has turned science away from the materialistic trend it had during the nineteenth century."[141] In a study of the relationship between complementarity and religious thought, "Physical Science and the Study of Religion," Bohr wrote that the renunciation of strict determinism and causality in quantum mechanics is directly related to dogmas of religion.

> Not only has the deterministic description of physical events, once regarded as suggestive support of the idea of predestination, lost its unrestricted applicability by the elucidation of the conditions for the rational account of atomic phenomena, but it must even be

[140] For minutes of these conferences, see *Zygon* 20 (1985): 107–220.

[141] W. Heisenberg, *Physics and Philosophy* (Harper, New York, 1958), p. 53.

> realized that mechanistic and finalistic argumentation, each within its proper limits, present inherently complementary approaches to the objective description of the phenomena of organic life. Moreover, the problem of free will, so pertinent in the philosophy of religions, has received a new background by the recognition, in modern psychology, of the frustration of attempts to order experience regarding our own consciousness as a causal chain of events, originally suggested by the mechanical conception of nature.[142]

It should be mentioned incidentally that there exists quite an extensive literature on the role of Bohr's complementarity principle as a heuristic device for theological studies as well as on its incorporation into theological thought.[143] According to Robert J. Russell, striking examples of its incorporation are the Chalcedonian formulation of the duality of the human and the divine in hypostatic union, Karl Barth's analysis of the perfections of God, Dietrich Bonhoeffer's insistence on the interwoven roles of belief and obedience, and Hans Küng's interpretation of the Resurrection.

[142] N. Bohr, "Physical Science and the Study of Religion," in *Studia Orientalia Ioanni Pedersen dicata* (E. Munksgaard, Copenhagen, 1953), pp. 385–390.

[143] D. M. MacKay, "'Complementarity' in scientific and theological thinking," *Zygon* 9 (1974): 225–244. C. B. Kaiser, "Christology and Complementarity," *Religious Studies* 12 (1976): 37–48. R. J. Russell, "Quantum Physics in Philosophical and Theological Perspective," in R. J. Russell, W. R. Stoeger, and G. V. Coyne, eds., *Physics, Philosophy, and Theology: A Common Quest for Understanding* (Vatican Observatory, Vatican City State, 1988), pp. 343–374. K. H. Reich, "The Relation between Science and Theology: The Case for Complementarity Revisited," *Zygon* 25 (1990): 369–390.

As discussed above, in his Credo, Einstein denied free will because of its incompatibility with determinism. The physicist Charles Alfred Coulson suggested that Einstein's conclusion could be avoided by resorting to Bohr's complementarity principle. Just as the type of observation determines whether a quantum object is a particle or a wave, so, according to Coulson, the point of view determines whether we accept free will or determinism. Mindful, probably, of Bohr's words "that the new situation in physics has so forcibly reminded us of the old truth that we are both onlookers and actors in the great drama of existence," Coulson argued, "Observed from without, the will is causally determined; observed from within it is free. The difference lies in the point of view for no answer at all can be given until we have specified explicitly the viewpoint of our observation, and said whether we are actor or spectator."[144] According to Coulson, who was also a Methodist lay preacher and spoke so much about the relation between physics and religion that he was called "the holder of the Chair of Theological Physics," religion and physics provide complementary perspectives of the same reality.

In general, however, scientists who believe that physics offers evidence for the existence of God derive their belief merely from the indeterminism of the quantum theory. The renowned physicist Arthur Stanley Eddington wrote that the conclusion to be drawn from examining the development of modern science is "that religion first became possible for a reasonable scientific man about 1927," the year of "the final overthrow of strict causality by Heisenberg,

[144] N. Bohr, *Atomic Theory and the Description of Nature* (Cambridge University Press, Cambridge, 1934), p. 119.

C. A. Coulson, *Science and Christian Belief* (Oxford University Press, London, 1955), p. 75.

Bohr, Born and others."[145] In the same vein, Edward Neville da Costa Andrade, Quain Professor of Physics at the University of London, declared in a radio talk that "the electron leads us to the doorway of religion."[146] The Oak Ridge nuclear physicist and Episcopal priest William Grosvenor Pollard said he was convinced that the providential action of God in every quantum event—which itself would be left undetermined—determines which among the possible options in accordance with quantum mechanics actualizes the final outcome.[147]

All these examples refer to advocates of the standard version of quantum mechanics. A proponent of the hidden-variable interpretation, the Dutch-American physicist Frederik Josef Belinfante stated,

> If I get the impression that nature itself makes the decisive choice what possibility to realize, where quantum theory says that more than one outcome is possible, then I am ascribing personality to nature, that is, to something that is always everywhere. Omnipresent eternal personality which is omnipotent in taking the decisions that are left undetermined by physical law is exactly what in the language of religion is called God.[148]

In other words, the outcome of every single quantum event, which is not uniquely determined by physical laws, is determined, according to Belinfante, by a supernatural force,

[145] A. S. Eddington, *The Nature of the Physical World* (Cambridge University Press, Cambridge, 1928), p. 350.

[146] Talk by E. N. da Costa Andrade, *Listener* 37 (July 10, 1947): 134.

[147] W. G. Pollard, *Chance and Providence* (C. Scribner, New York, 1958).

[148] F. J. Belinfante, *Measurement and Time Reversal in Objective Quantum Theory* (Pergamon Press, Oxford, 1975), pp. 98–99.

God. In view of the fact that Belinfante's main interest in the seventies focused on hidden-variable formulations of quantum mechanics, it seems not unreasonable to interpret his statements as a declaration of an apothesis of hidden variables.[149]

Theistic explanations of the quantum-mechanical indeterminism, as those suggested by Belinfante and others, would have been unacceptable to Einstein, not only because of his rejection of hidden variables, but primarily because such an approach would have been for him what is called a doctrine of "God of the gaps," that is, a hypothesis to account for scientifically unexplained phenomena in terms of theistic conceptions.[150] As he stated in his 1940 address to the New York Conference on Science, Philosophy and Religion, such doctrines could never be refuted by science, because they always take refuge in those domains in which scientific knowledge has not yet been able to set foot; but to profess such a doctrine "would not only be unworthy, but also fatal. For a doctrine which is able to maintain itself not in clear light but only in the dark, will of necessity lose its effect on mankind, with incalculable harm to human progress."[151]

Much has been written about Einstein's views on quantum mechanics, how he thought the standard version should be "completed" without resort to hidden vari-

[149] F. J. Belinfante, *A Survey of Hidden Variables* (Pergamon Press, Oxford, 1973).

[150] Belinfante's view was shared, e.g., by the plasma physicist Shoici Yoshikawa, who declared that "the hidden variables of quantum mechanics are under God's power." See H. Margenau and R. A. Varghese, eds., *Cosmos, Bios, Theos* (Open Court, La Salle, Ill., 1992), p. 133.

[151] A. Einstein, *Science and Religion.*

ables; whether he favored a statistical interpretation according to which the wave function describes not a single system but an ensemble of systems; whether, in particular, what Arthur Fine calls the "prism-model approach," could resolve all the epistemological problems Einstein has raised; or finally whether Einstein believed that probably only a multidimensional unified theory could provide a satisfactory answer.[152]

Einstein himself has never explicitly announced a full-fledged and, for him, satisfactory formalism of the theory of quanta. If no "Einsteinian quantum mechanics" exists, one can hardly speak of its theological significance or its theological implications. It is, however, not illogical to study the theological connotations, if there are any, of Einstein's denial of orthodox quantum mechanics.

In a thought-provoking article published on the occasion of the Einstein Centennial in 1979, Gunther Siegmund Stent contended that Einstein's opposition to Bohr's interpretation of quantum mechanics had a profound theological significance. In Stent's view, the Einstein-Bohr debate reflected ultimately a clash between two opposing religious worldviews. "Einstein represented the traditional monotheistic viewpoint of Western science, whereas Bohr showed what it takes to be a great scientist and yet pay more than lip-service to atheism."[153] Just like "the archaeologist facing Stonehenge" is convinced that the huge stones are placed in their circular pattern not by chance or whimsey, but by design, so the scientist, who believes in

[152] A. Fine, *The Shaky Game* (University of Chicago Press, Chicago, 1986).

[153] G. S. Stent, "Does God Play Dice?" *The Sciences 19* (March 1979), 18–21.

monotheism, sees his work as an attempt "to fathom the Creator's intentions and design." To prove that this was indeed Einstein's point of view, Stent recalled that, whenever a certain theory appeared to Einstein arbitrary or forced, he said "God does not do anything like that" or "Subtle is the Lord, but malicious He is not" [Raffiniert ist der Herrgott aber boshaft ist er nicht]. Stent was apparently unaware of Einstein's statements that were even more to the point, such as "I wish to know how God created this world . . . the rest are details." Stent declared that unlike Einstein, Bohr did

> not regard the world as an objective reality with a given structure (let alone 'design') conceptually separable from us as observers. Instead the world is simply there, with us in it as an integral and inseparable part. Thus, there must be limits to the depth of understanding that we can hope to gain of the world, both because of our joint role as spectators and actors in the drama of existence and because that drama, lacking an author, has no plot.

Stent's argument hinges to a large extent on the assumption that Einstein's frequent references to God have been made in their literal sense and not merely in a manner of speaking. Stent also pointed out that Bohr's thinking conforms more with the philosophies of the Far East than with the Judeo-Christian tradition and that "Bohr was probably the first great modern scientist to hold such views."

In fact, as early as 1937, Bohr declared in his Bologna lecture that to find a parallel to the conceptual structure of quantum mechanics we have to turn "to that kind of epistemological problems with which already Buddha and

Lao-tse have been confronted, when trying to harmonize our position as spectators and actors in the great drama of existence."[154] When Bohr accepted knighthood in 1947, he chose as the emblem for his coat of arms the yin-yang, which in the ancient Chinese religion symbolized the unification of opposites by the complementarity of opposing principles or forces in nature.

Erwin Schrödinger was deeply interested in ancient Oriental religious doctrines and, in particular, in the Vedanta as his essays "The Vedantic Vision" and "The doctrine of identity: light and shadow" clearly indicate.[155] It has even been contended that he "got his cat paradox from the Sankhya," a very old and pessimistic doctrine of Hinduism.[156] David Bohm, who was greatly influenced by Einstein, admitted repeatedly that he was also inspired by the Indian spiritual leader Jiddu Krishnamurti, especially with respect to what Bohm called "the fragmentary Western approach" to epistemological and ontological problems in physics.[157] Bohm's notion of the implicate order, according to which any element contains enfolded within itself the totality of the universe, and his idea of the world as an unfolding-enfolding movement of transformations undoubtedly reveal some affinity with Eastern philosophical and religious thought.[158]

[154] N. Bohr, *Atomic Physics and Human Knowledge* (Wiley, New York, 1958), p. 20.

[155] E. Schrödinger, *My View of the World* (Cambridge University Press, Cambridge, 1964), pp. 18–22, pp. 92–104.

[156] D. Harrison, "Teaching the Tao of Physics," *American Journal of Physics* 47 (1979): 779–783.

[157] See D. Bohm, *Quantum Theory*.

[158] D. Bohm, *Wholeness and the Implicate Order* (Routledge and Kegan Paul, London, 1980). For details on the influence of Oriental thought

Unlike Bohr, Schrödinger, and Bohm, Einstein never showed any interest in Far Eastern philosophy and never expressed any sympathy with Oriental religious thought or mysticism. On the contrary, he condemned "the mystical trend of our time . . . as a symptom of weakness and confusion."[159] Yet, Einstein nevertheless played an important, though only indirect and certainly unwanted, role in the proliferation of a literature that relates modern physical conceptions with Far Eastern religious thought and claims that "the general spirit of the two disciplines is the same."[160] The above-outlined conceptual development initiated by Einstein led from the EPR incompleteness argument through Bell, Aspect, and others to the refutation of Einstein's local realism and to the result that there exist instantaneous noncausal correlations between spatially, or even spacelike, separated physical systems. These telepathy-like relations, as Einstein has called them, suggested that ultimate reality should be attributed not to the physical objects themselves but rather to their relations with other objects. In other words, relations are more fundamental than the relata, and the essence of an object is its connectedness and dependency within a complicated web of interrelations.

Such ideas can be found in ancient sacred writings of the

---

on Bohm, see J. A. Schumacher and R. M. Anderson, "In Defense of Mystical Science," *Philosophy East and West* 29 (1979): 73–90. R. J. Russell, "The Physics of David Bohm and Its Relevance to Philosophy and Theology," *Zygon* 20 (June 1985): 135–158.

[159] "Der mystische Zug unserer Zeit . . . ist für mich nur ein Symptom von Schwäche und Zerfahrenheit." Einstein to L. Halpern-Neuda, 5 February 1921. Einstein Archive, reel 43-847.

[160] F. Capra, "The Dance of Shiva," *Main Currents of Modern Thought* 29 (September–October 1972): 15–20.

Far East. The Buddhistic Madhyamika ("Middle Way") doctrine teaches the vanity or "emptiness" of the ego and extends this "vacuity" ("sunyatavada") to the *dharmas*, the elements of things. The Vedas, the canonical scriptures which in Hinduism are regarded as *sruti* (revealed words of divinity), proclaim the unity of the universe and conceive separability as merely an illusion, not unlike the Eleatic doctrine that all beings constitute one single being.

Such analogies prompted, to mention just one example, the physicist Victor Mansfield to "compare the pivotal doctrine of emptiness with the current view of incompleteness and nonlocality from physics and then apply Madhyamika principles to their understanding." "I will show," he continued, "that there is an intimate relationship among the Madhyamika conception of emptiness, the experimental refutation of Bell's Inequality, and the fundamental principles of quantum mechanics."[161] Mansfield justified his analysis of this relationship on the grounds that "while Madhyamika may help in understanding quantum mechanics, quantum mechanics may also help in understanding Madhyamika." He wisely added that he "in no way attempts to prove the validity of Madhyamika Buddhism through physics."

The advice not to use physics to validate a religious doctrine, or to use religious dogma to validate a physical theory, has not always been heeded by the authors of books that relate quantum mechanics with Far Eastern religious thought. Although almost all of them refer to the EPR argument and its consequences and thus, implicitly at least,

[161] V. Mansfield, "Madhyamika Buddhism and Quantum Mechanics: Beginning a Dialogue," *International Philosophical Quarterly* 29 (1989): 371–391.

acknowledge their indebtedness to Einstein, I do not discuss these books in any detail. I only quote the titles of the most popular publications of this genre in the notes and add that Fritjof Capra's *The Tao of Physics* became a best-seller with more than a million copies sold and Gary Zukav's *The Dancing Wu-Li Masters* won an American Book Award in 1980.[162]

These publications have aroused quite a number of critical responses, some mildly uncomplimentary, some rather devastating. The main issue concerns whether the similarities claimed to exist between Eastern religions and quantum mechanics are what has been aptly called "parallels of analogy" or "parallels of identity," that is, whether they are intended to refer to different domains of objects and different levels of reality or to the same identity of objects.[163] The critics admit that there is no harm in drawing analogies between different fields of human thought and that such analogies may have a didactic and even

[162] R. G. H. Siu, *The Tao of Science* (MIT Press, Cambridge Mass., 1957). F. Capra, *The Tao of Physics* (Shambhala, Berkeley; Oxford University Press, Oxford, 1975, 1983); *The Turning Point* (Simon and Schuster, New York, 1982). G. Zukav, *The Dancing Wu-Li Masters* (W. Morrow, New York; Hutchinson, London, 1979). A. de Riencourt, *The Eye of Shiva: Eastern Mysticism and Science* (Souvenir Press, London, 1980). M. Talbot, *Beyond the Quantum* (Bantam, New York, 1981); *Mysticism and the New Physics* (Viking Press, New York, 1993). R. Weber, *Dialogues with Scientists and Sages: The Search for Unity* (Routledge and Kegan Paul, London, 1986). M. B. Towsey, *Eternal Dance of Macrocosm* (Proutist Publishers, Sydney, 1986). F. A. Wolf, *Taking the Quantum Leap* (Harper, New York, 1989).

[163] A. Balasubramaniam, "Explaining Strange Parallels: The Case of Quantum Mechanics and Madhyamika Buddhism," *International Philosophical Quarterly* 32 (1992): 205–223.

heuristic value. What they object to is that the similarities quoted by the authors of those books are regarded by these authors as "parallels of identity" and not merely as "parallels of analogy." For they attempt to explain, justify, or validate Eastern religious mysticism on the grounds of its similarities with quantum mechanical conceptions. (For details on further critical comments, see the essays listed in note 164.)[164]

The literature on quantum mechanics and Eastern religious thought appeared (see note 162) only after Einstein's death in 1955. As to how Einstein would have reacted, had he been alive and read these publications, the most likely answer is that he would not have been very hospitable to these ideas, if only because they were based on the orthodox version of quantum mechanics and the denial of local realism.

A similar question may be asked with respect to a by-product of the general theory of relativity that led, after Einstein's death, to the claim that physics and religion are not only compatible but also support or even validate and

[164] S. P. Restivo, "Parallels and Paradoxes in Modern Physics and Eastern Mysticism," *Social Studies of Science* 8 (1978): 143–181; 12 (1982): 37–71; *The Social Relations of Physics, Mysticism, and Mathematics* (Reidel, Dordrecht, 1983). C. Clarke, "Comment: On Physics and Mysticism," *Theoria to Theory* 14 (1981): 333–337. D. H. Esbenshade, Jr., "Relating Mystical Concepts to Those of Physics: Some Concerns," *American Journal of Physics* 50 (1982): 224–228. R. P. Crease and C. C. Mann, "Physics for Mystics," *The Sciences* 27 (1987): 50–57. E. R. Scerri, "Eastern Mysticism and the Alleged Parallels with Physics," *American Journal of Physics* 57 (1989): 687–692. R. K. Clifton and M. G. Regehr, "Toward a Sound Perspective on Modern Physics: Capra's Popularization of Mysticism and Theological Approaches Reexamined," *Zygon* 25 (1990): 73–104.

confirm each other. This is the Big Bang cosmology, a class of cosmological models that assert that the universe expanded to its present state from an initial singularity that occurred some 15 billion years ago.

As STATED at the end of chapter 1, Einstein, in his 1917 essay, "Cosmological Considerations," raised the status of cosmology from a jumble of mythical speculations to a respectable scientific discipline.[165] Hence, any cosmological theory using the mathematical apparatus of general relativity owes a great debt to Einstein. This applies also to the various versions of the Big Bang theory. The Big Bang cosmology is based on Aleksandr Friedman's 1922 nonstationary solutions of Einstein's field equations, on Edwin Hubble's 1929 observational discovery of the expansion of the universe, and on George Gamow's 1946 modification of Georges Lemaître's hypothesis of the exploding "primeval atom" containing all the matter of the universe. It explains the physical process of the evolution of the universe to its present state by resorting to the generally accepted so-called standard model of the theory of elementary particles.[166] Finally, as far as it refers to the very beginning of the universe, it generally applies the concept of a singularity, a point in space-time at which the space-time curvature is infinite. In fact, according to the 1970 singularity theorem of Stephen W. Hawking and Roger Penrose, such a singularity is a necessary consequence of extrapolating, back-

[165] A. Einstein, "Cosmological Considerations on the General Theory of Relativity."

[166] S. Weinberg, *The First Three Minutes* (Basic Books, New York, 1977).

ward in time, any cosmological model, whether homogeneous or not, if timelike or geodesic incompleteness is used as the indication of space-time singularities.[167]

Einstein probably would have accepted these developments insofar as they are grounded on empirically well-established facts. His acceptance would have been reluctant because they do not conform with his conviction that all physical phenomena ultimately should be accounted for in terms of continuous fields everywhere. This conviction probably motivated him in 1922 to reject Friedman's solution because it implied the existence of a "singular point" in the space-time continuum. Nevertheless, in 1923, he accepted the nonstationary solution and even called it "correct and clarifying" [richtig und aufklärend].[168] In a 1927 paper, coauthored with Jakob Grommer, Einstein admitted the possibility of "singular points" when he declared that "all attempts of the last years to explain the elementary particles of matter in terms of continuous fields have failed," and "one is therefore forced into the direction of conceiving elementary particles as singular points."[169] In later years, his attitude on this issue, especially with respect to cosmological problems, is best characterized by the following statement.

[167] S. W. Hawking and R. Penrose, "The Singularities of Gravitational Collapse and Cosmology," *Proceedings of the Royal Society, London, A 314* (1970): 529–548.

[168] A. Einstein, "Bemerkung zu der Arbeit von A. Friedman: Über die Krümung des Raumes," *Zeitschrift für Physik* 11 (1922): 326; "Notiz zu der Bemerkung zu der Arbeit von A. Friedman: Über die Krümmung des Raumes," *ibid.*, 16 (1923): 228.

[169] A. Einstein and J. Grommer, "Allgemeine Relativitätstheorie und Bewegungsgesetz," *Sitzungsberichte der Preussischen Akademie der Wissenschaften* 1927 (1927): 2–12.

> For large densities of field and matter, the field equations and even the field variables which enter into them have no real significance. One may not therefore assume the validity of the equations for very high density of field and matter, and one may not conclude that the "beginning of the expansion" must mean a singularity in the mathematical sense. All we have to realize is that the equations may not be continued over such regions.[170]

It is, of course, precisely the very "beginning of the expansion," the question of how and why the universe started, that attracted the attention of theologians and philosophers.

Some variants of the Big Bang theory are intimately related to the quantum theory. When the space-time dimensions of the primeval universe were still very small, not larger than the Planck length of 10^{-33} cm and the Planck time of 10^{-43} sec, quantum effects must have certainly played an important role. In this context in 1973, Edward P. Tryon proposed a Big Bang model stating that our universe originated as a quantum fluctuation of the vacuum, a common phenomenon in quantum field theory. Applying Einstein's mass-energy relation and an argument similar to that used by Margenau, namely that the total energy of the universe is zero, Tryon pointed out that the Heisenberg indeterminacy relation between energy and time implies that our universe could exist for eternity.[171] On the philosophical question of why there is something rather than nothing or why such a fluctuation occurred, Tryon replied that "our

[170] A. Einstein, "On the 'Cosmological Problem,'" Appendix I in *The Meaning of Relativity* (Methuen, London, 1950), p. 123.

[171] H. Margenau, *Thomas and the Physics of 1958.*

universe is simply one of those things which happen from time to time."[172] Tryon's quantum-cosmological proposal of a *creatio ex nihilo* has been elaborated and modified by a number of theoreticians.[173] The question has been raised as to whether it would be possible to create a "universe" in a well-equipped laboratory. Such a possibility has been seriously investigated by Edward Farhi and Alan H. Guth. They concluded that "to create conditions in a small region of space which would give rise to a new universe . . . would require an energy density that is far too high to be provided by any known technology."[174]

Until the mid-sixties, the Big Bang theory had to compete with the Steady-State theory proposed, in the late forties, by Hermann Bondi and Thomas Gold and most thoroughly elaborated by Fred Hoyle, who incidentally also coined the term "Big Bang," though merely as an expression of derision. Hoyle showed that a modification of the mathematical structure of the energy-momentum tensor in Einstein's field equations makes it possible to reconcile the expansion of the universe with the perfect cosmological principle, according to which the average properties of the universe are the same everywhere and at all times. In particular, the average local density of the universe remains constant because of the exis-

[172] E. P. Tryon, "Is the Universe a Vacuum Fluctuation?" *Nature* 246 (1973): 396–397; "What Made the World?" *New Scientist* 101 (8 March 1984): 14–16.

[173] A. Brout, F. Englert, and E. Gunzig, "The Creation of the Universe as a Quantum Phenomenon," *Annals of Physics* 115 (1978): 78–106. D. Atzkin and H. Pagels, "Origin of the Universe as a Quantum Tunneling Event," *Physical Review D* 25 (1982): 2065–2073. A. Vilenkin, "Creation of Universes from Nothing," *Physics Letters 117* B (1982): 25–28.

[174] E. Farhi and A. Guth, "An Obstacle to Creating a Universe in the Laboratory," *Physics Letters 183 B* (1987): 149–155.

tence of a "creation field" that continually produces new matter in the form of hydrogen atoms that combine and give birth to new stars and galaxies. Because this field carries negative energy that compensates for the positive energy of the created matter in accordance with Einstein's mass-energy relation, the energy conservation law need not be violated. Furthermore, by claiming that the universe had no beginning, the Steady-State theory, unlike the Big Bang theory, avoids the conceptual problem raised by a time-singularity beyond which the history of the universe cannot be traced. "It is against the spirit of scientific enquiry to regard observable effects as arising from 'causes unknown to science,' and this in principle is what creation-in-the-past implies."[175]

The debate between the two rival theories came to an end in 1964 when Arno Penzias and Robert Wilson accidentally discovered the 2.7° K cosmic microwave background radiation that Gamow had predicted to exist as a relic of the initial explosion. The Big Bang theory gained additional support when astronomical deep-space observations showed that the universe must have been different in the past from what it is now. More recently, in April 1992, the Cosmic Background Explorer (COBE) verified the existence of ripples in the background radiation, a necessary precondition for the formation of galaxies, granting the Big Bang theory, in one variant or another, almost universal acceptance.

Discussing these variants, among them, the important inflationary models, would lead us too far from the topic

[175] F. Hoyle, "A New Model for the Expanding Universe," *Monthly Notices of the Royal Astronomical Society* 108 (1948): 372–382.

of this chapter. Precisely because of its theological implications, one exception deserves attention: the quantum gravitational model proposed by Stephen Hawking in collaboration with James B. Hartle and published in 1983. In this proposal, as Carl Sagan put it, "Hawking embarks on a quest to answer Einstein's famous question about whether God had any choice in creating the universe."[176]

The following sketch gives a nontechnical account of Hawking's reasoning. Making use of the path-integral approach to quantum mechanics and defining time, in the early stage of the universe, as an internal parameter in terms of some property of the universe such as its mass-energy density, Hawking made it redundant to add an external time coordinate to the early stage of the universe's evolution. In order to visualize Hawking's model, recall that, according to conventional Big Bang theory, the expanding universe, if represented in a Minkowski-like four-dimensional space-time diagram, resembles an ice cream cone whose pointlike tip is the initial singularity. Hawking showed that the near region of this tip can be modified so that its time dimension gradually turns into a space dimension and the tip becomes a hemisphere as one moves beyond the Planck time. At the bottom of the hemisphere, the four-dimensional space-time degenerates into a four-dimensional spatial manifold from which time gradually emerges without an abrupt coming into existence, so that here is no instantaneous beginning of time. Thus, "one would have solved the problem of the initial boundary conditions of the Universe: the boundary conditions are

[176] C. Sagan, Intoduction to S. W. Hawking, *A Brief History of Time* (Bantam, Toronto, 1988), p. x.

that it has no boundary."[177] Or as Hawking phrased it in his best-seller *A Brief History of Time*, "There would be no singularities at which the laws of science broke down and no edge of space-time at which one would have to appeal to God or some new law to set the boundary conditions for space-time. . . . The Universe would be completely self-contained and not affected by anything outside itself. It would neither be created nor destroyed. It would just BE." Hawking thus came to the following conclusion, "So long as the universe had a beginning, we could suppose it had a creator. But if the universe is really completely self-contained, having no boundary or edge, it would have neither beginning nor end: it would simply be. What place, then, for a creator?"[178]

The critical reader will probably raise the question of whether Hawking's singularity-free quantum gravitational model does not violate the above-mentioned Hawking-Penrose singularity theorem. Anticipating such a question Hawking pointed out that

> What the singularity theorems really indicate is that the gravitational field becomes so strong that quantum gravitational effects become important: classical theory [based on Einstein's general theory] is no longer a good description of the universe. So one has to use a quantum theory of gravity to discuss the very early stages of the universe. . . . It is possible in the quantum theory for the ordinary laws of science to hold

[177] J. B. Hartle and S. W. Hawking, "Wave Function of the Universe," *Physical Review D* 28 (1983): 2960–2975.

[178] S. W. Hawking, *A Brief History of Time*, p. 136, pp. 140–141.

> everywhere, including at the beginning of time; it is not necessary to postulate new laws for singularities, because there need not be any singularities in the quantum theory.[179]

HAVING REVIEWED a number of cosmological theories based on Einstein's general relativity, I now discuss their implications, if any, for theological thought. As far as Einstein himself and his general theory are concerned, he emphatically denied any relation between his theory and theology.[180] In 1917 when he applied the theory to cosmological problems and introduced the cosmological constant in order to avoid a nonstatic solution, Spinoza's influence on his thinking about cosmology could be recognized.[181] At that time, Einstein thought, though never stated explicitly, that in the absence of matter and energy—or more precisely, if the energy-momentum tensor T_{mn} in the field equations $G_{mn} = k\, T_{mn}$ is zero—the equations have no possible solution.[182] He believed that this means that neither space nor time exists in a universe devoid of matter. "In my opinion it would be unsatisfactory," he wrote on March 24, 1917, to Willem de Sitter, "if a world without matter could conceivably exist. The g_{mn}-field should rather be *determined by matter and*

[179] Ibid., p. 133.

[180] Eddington, *The Philosophy of Physical Science*, p. 7.

[181] "Deum, sive omnia Dei attributa esse immutabilia," *Ethics*, col. 2 to proposition 20, pt. I; "rem extensam (et rem cogitantem) Dei attributa esse," ibid., col. 2 to proposition 14. See also J. A. Wheeler, "Beyond the Black Hole," in H. Woolf, ed., *Some Strangeness in Proportion* (Addison-Wesley, Reading, Mass., 1980), p. 354.

[182] In these equations, the tensor G_{mn} determines the curvature of the space-time and k is a constant of proportionality.

should be incapable of existing without it. This is the core of what I understand by the requirement for the relativity of inertia."[183]

Einstein's non-Newtonian idea that time cannot exist without the presence of matter has a long history of its own. It can be traced back to Plato, who described in his sole dialogue on science, the *Timaeus,* how the Demiurge created the world and endowed it with a "moving image of eternity" that manifests itself in the motions of the heavenly bodies. Thus, "simultaneously with the Heaven He contrived the production of days and nights and months and years, which existed not before the Heaven came into being."[184] Fascinated by Plato's *Timaeus,* Philo Judaeus (Philon the Jew) applied this idea to his exegesis of the first words of *Genesis*: "In the beginning God created the heaven and the earth." "Beginning," he wrote, "should not be conceived in a chronological sense," for "time cannot be before there was a world. Time began

[183] "Es wäre nach meiner Meinung unbefriedigend, wenn es eine denkbare Welt ohne Materie gäbe. Das g_{mn}-Feld soll vielmehr durch die Materie bedingt sein, ohne dieselbe nicht bestehen können. Das ist der Kern dessen, was ich unter der Forderung von der Relativität der Trägheit verstehe." Einstein to W. de Sitter, 24 March 1917, Einstein Archive, reel 20-548. Einstein's hypothesis "no matter—no space-time" was soon shown to be mistaken by de Sitter in his article "On the Relativity of Inertia: Remarks Concerning Einstein's Latest Hypothesis," *Proceedings of the Section of Sciences, Koninklijke Akademie van Wettenschappen te Amsterdam* 19 (1917): 1217–1225. De Sitter's solution of the field equations for an empty and static universe was the second general relativistic cosmological model; Einstein's 1917 model was the first.

[184] Plato, *Timaeus E 7* (Heinemann, London; Putnam's Sons, New York, 1929; Harvard University Press, Cambridge, Mass., Heinemann, London, 1942), p. 77.

either simultaneously with the world or after it. For time is a measured space determined of the world's movement, and since movement could not be prior to the object moving . . . it follows of necessity that time also is either coeval with or later born than the world."[185] By interpreting Scripture in terms of the physical knowledge available at that time, Philo tried to show that Greek science as taught in Alexandria of the first century A.D. had its origin in the teaching of Moses.

Similar apologetic exegeses of *Genesis* and the Bible in general can be found in the writings of Origen, the head of the famous Catechetical School of Alexandria in the early third century, especially in his *De Principiis,* one of the earliest treatises on systematic theology. His disciple Saint Basil of Caesarea likewise interwove scientific references in his theological writings and homilies, such as in his *On the Hexameron,* a commentary on the six days of creation. Support for the authority of Christian doctrine from physical explanations was also a favorite theme in the six books of the *Exameron* written by Ambrose, bishop of Milan. Baptized by Ambrose, Augustine, the bishop of Hippo, is probably the best known exegete of *Genesis* as far as the notion of time is concerned. Like Philo, Augustine was strongly influenced by Plato, whose *Timaeus* he read in a Latin translation. Augustine declared, "Non in tempore sed cum tempore finxit Deus mundum."[186] Au-

[185] Philo, *On the Account of the World's Creation given by Moses (De Opificio Mundi)* (Heinemann, London; Putnam's Sons, New York, 1929), pp. 20–21. See also Philo, *Allegorical Interpretation of Genesis* II, III, ibid., pp. 146–473.

[186] "Not in time but with time God created the world." Augustine, *De Civitate Dei,* book 11, 5.

gustine wrote at least five, not all complete, exegetical treatises about the beginning of the Book of Genesis.[187] The two most important of these have recently been published in an English translation.[188] In one passage, Augustine defended Genesis 1:1 against critics who ask, "If God made heaven and earth in some beginning of time, what was he doing before he made heaven and earth? And why did he suddenly decide to make what he had not previously made through eternal time?" Unable to provide a scientific answer, Augustine, obviously referring to John's words "In the beginning was the Word" (John 1: 1, 3), replied: "God made heaven and earth in the beginning, not in the beginning of time, but in Christ. For he was the Word with the Father, through whom and in whom all things were made."[189] Modern cosmology faces similar questions, though generally formulated in different terms. For example, why did the Big Bang start about 15 billion years ago and not earlier or later?

Because of its comprehensive synthetic character, the Aristotelian philosophical system gained dominant influence on theological thought in the later Middle Ages. When it had to be assimilated to harmonize with traditional religious doctrine, the Aristotelian thesis that "the world as a whole was not generated and cannot be destroyed" raised a serious problem.[190] For example, faithful

[187] G. Pelland, *Cinq Études d'Augustin sur le Début de la Genèse* (Declée, Paris; Bellarmin, Montreal, 1972).

[188] Augustine, *On Genesis—Two Books on Genesis against the Manichees and On the Literal Interpretation of Genesis: An Unfinished Book* (Catholic University of America Press, Washington, D.C., 1991).

[189] Ibid., p. 49.

[190] Aristotle, *De Caelo* (On the Heavens), book 2, chap. 1 (283 b 27); *The Physics*, book 8, chap. 1 (250 b 11–252 b 6).

to the biblical tradition but strongly influenced by Aristotle, Moses Maimonides, tried in vain to find a logically acceptable compromise.[191] The Fourth Lateran Council, convened in November 1215 by Innocent III in Rome, proclaimed the thesis of the beginning of the world in time as an article of faith. Saint Thomas Aquinas also declared, "We hold by faith alone, and it cannot be proved by demonstration, that the world did not always exist . . . the reason being that the newness of the world cannot be demonstrated by the world itself."[192]

Understandably, such statements by leading authorities had the effect of considerably lessening interest in the following centuries in the hexaemeron literature, which had been intended to offer a scientific explanation of the first book of Genesis. In fact, in the early nineteenth century, a prominent theologian declared it "a lost effort to try a physical demonstration of the account of creation as given by Moses."[193] However, with the advent of the Big Bang cosmology, which stated that the universe originated at a definite time in the past with a unique event that could easily be interpreted as a divine act of creation,

[191] M. Maimonides, *The Guide of the Perplexed* (University of Chicago Press, Chicago, 1963), book 2, chap. 30. See also A. Nuriel, "The Question of a Created or Primordial World in Maimonides," *Tarbiz* 33 (1964): 372–387. S. Klein-Braslavy, "Creation of the World and Maimonides' Interpretation of Genesis I–V," in S. Pines and Y. Yovel, *Maimonides and Philosophy* (M. Nijhoff, Dordrecht, 1986), pp. 65–78.

[192] "Quod mundum non semper fuisse sola fide tenetur." *Divi Thomae Aquinatis Summa Theologica* (Senatores, Rome, 1886), question 46, article 2, p. 377; *Summa Theologica* (Great Books of the Western World, vol. 19) (Encyclopaedia Britannica, Chicago, 1952), pp. 250–253.

[193] K. G. Bretschneider, *Handbuch der Dogmatik* (Barth, Leipzig, 1814, 1828), p. 587.

the hexaemeric interpretations experienced a widespread revival. The authors of the modern hexaemerons, which differed both in substance and in style from their medieval prototypes, were encouraged in their work by the clerical hierarchy. For example, in an address delivered in November 1951 to the Pontifical Academy of Science, Pope Pius XII approved for the first time *ex cathedra* interpretation of the text of Genesis as an allegorical narrative adapted to the understanding of the common folk. He praised the cosmologists for having demonstrated that the universe is "the work of creative omnipotence, whose power, set in motion by the mighty *Fiat* pronounced billions of years ago by the Creating Spirit, spread out over the universe."[194]

The November 1979 meeting of the Pontifical Academy, held in honor of the centenary of the birth of Einstein and presided over by Pope John Paul II, also deserves mention. After the opening talk by Carlos Chagas, the president of the Academy, on Einstein's life and his philosophy of nature, and Paul A. M. Dirac's slightly more technical report on Einstein's contributions to physics, Victor F. Weisskopf explicitly referred to the Big Bang cosmology in his address. "The modern view that the universe originated from an infinitely compressed hot assembly of primal matter in the big bang and the subsequent expansion of the universe," Weisskopf declared, "are ideas that were spawned by Einstein's conception of space and time." In his concluding address, the Pope expressed his wish "to render to Albert Einstein the honor that is due him for the eminent

[194] "Science and the Catholic Church," *Bulletin of the Atomic Scientists* 8 (1952): 143–146, 165.

contribution he has made to the progress of science—that is, to the knowledge of the truth present in the mystery of the universe." The Pope continued:

> On the occasion of this solemn commemoration of Einstein, I would like to confirm again the Council's declaration on the autonomy of science in its function of searching for the truth inscribed during the creation by the finger of God. Filled with admiration for the genius of the great scientist, in whom is revealed the imprint of the creative spirit, without intervening in any way with a judgment on the doctrines concerning the great systems of the universe, which is not in her power to make, the Church nevertheless recommends these doctrines for consideration by theologians in order to discover the harmony that exists between scientific truth and revealed truth.[195]

It was not only the Church that showed her willingness to repent her former attitude, as shown in the case of Galileo, and to improve her relations with the cosmologists. Because of the insights gained from the Big Bang theory, the cosmologists began to find a common language with theology. A vivid description of this process has been given by the astronomer Robert Jastrow.

> Now we see how astronomical evidence leads to the biblical view of the origin of the world. The details differ, but the essential elements in the astronomical and biblical accounts of Genesis are the same: the

[195] "Einstein Session of the Pontifical Academy," *Science* 207 (1980): 1159–1167. See also S. Hawking's comments on an audience with the Pope in *A Brief History of Time*, pp. 116, 136.

> chain of events leading to man commenced suddenly and sharply at a definite moment of time, in a flash of light and energy. . . . For the scientist who has lived by the faith in the power of reason, the story ends like a bad dream. He has scaled the mountains of ignorance; he is about to conquer the highest peak; as he pulls himself over the final rock, he is greeted by a band of theologians who have been sitting there for centuries.[196]

This convergence between scientists and theologians led to the writing of numerous essays and books showing that scientific evidence agrees with the biblical account of the origin of the universe and that, moreover, the new cosmology provides an important tool for the understanding of the book of Genesis. Not a few of these modern versions of medieval hexaemerons have been written by theologians, such as Abbé Charles Hauret, who devoted themselves to the study of physics and astronomy.[197]

Most of the other articles on this subject were written by religious professional scientists such as physicists Henry Margenau, Gerald Schroeder, and Cyril Domb.[198] A widely

[196] R. Jastrow, *God and the Astronomers* (W. W. Norton, New York, 1978), p. 116.

[197] C. Hauret, *Origines de l'Universe et de l'Homme d'après la Bible* (Gabalda, Paris, 1950); *Beginnings: Genesis and Modern Science* (Priory Press, Dubuque, Ia., 1955, 1964). Other examples are H. Junker, *Die Biblische Urgeschichte* (Hanstein, Bonn, 1932); R. T. Murphy, *The First Week* (Rosary Press, Somerset, 1962); F. Ceuppens, *Genèse I–III* (Declée, Paris, 1945); Th. Schwegler, *Die biblische Urgeschichte im Licht der Natur- und Geisteswissenschaften* (Biblische Beiträge, Schweizerische Katholische Bibelbewegung, Heft 11), (Walter, Olten, 1951).

[198] See also P. W. Stoner, "Genesis I in the Light of Modern Astronomy," in J. C. Monsma, *The Evidence of God in an Expanding Universe*

known example is the monograph *In the Beginning* by the solid-state physicist Nathan Aviezer, who devoted a separate chapter to each of the six days of creation to demonstrate that the scientific evidence provided by the Big Bang theory is consistent with the biblical account in Genesis. The chapter dealing with the first day of creation opens with a brief review of the Big Bang cosmology and states that "the primeval fireball *is* the creation of the universe," and it interprets the biblical passage "Let there be light" as "designating the creation of the primeval fireball—the big bang—that signals the creation of the universe." The chapter on the first day ends with the summary, "In short, hundreds of years of intense scientific effort by some of the finest minds that ever lived has finally produced a picture of the universe that is in striking agreement with the simple words that appear in the opening passages of the book of Genesis."[199] As Aviezer stressed in a "Credo" at the end of his book, and as Maimonides and Thomas Aquinas had taught long ago, a person's religious commitment should not depend on the consistency between physics and the Bible. Aviezer probably considered it possible that the Big Bang theory, like every scientific theory, can, in spite of its numerous confirmations at present, be refuted one day in favor of an alternative theory which is less consistent with the biblical report. In fact,

(Putnam's Sons, New York, 1958), pp. 137–142; the essays by A. Radkowsky and A. Carmell in A. Carmell and C. Domb, eds., *Challenge* (Feldheim, Jerusalem, 1978), pp. 68–92, pp. 306–342; O. Gingerich, "Let There Be Light: Modern Cosmogony and Biblical Creation," in R. M. Frye, *Is God a Creationist?* (Scribner's Sons, New York, 1983), pp. 119–137.

[199] N. Aviezer, *In the Beginning—Biblical Creation and Science* (Ktav, Hoboken, N.J., 1990), p. 17. The book has been published also in French, Russian, Spanish, and Portuguese.

not every scientist now accepts the Big Bang theory. Thus, for example, John Maddox wrote in 1989 that this theory "is unlikely to survive the decade ahead . . . for it provides no explanation at present for quasars and the source of the known hidden mass in the Universe. It will be a surprise if it survives the Hubble telescope."[200] It should also be noted that alternative theories, apart from the Steady-State theory, have been proposed by competent cosmologists.[201]

It cannot be doubted that, psychologically speaking, the Big Bang theory lends some support to the theistic doctrine of the creation by divine fiat. If interpreted in the sense of whether the existence of the world depends on a definite choice of the laws of nature, as suggested by the proponents of the anthropic principle, Einstein's problem of "whether God could have made the world different" does not exclude the deistic assumption that the act of creation is a one-time historical event after which God no longer interacted with the world.[202] Because the Big Bang theory is also compatible with deism, it may be asked whether it is not erroneous to suppose, as is usually done, that the Big Bang cosmology is theologically preferable to the Steady-State cosmology, which denies the one-time event but as-

[200] J. Maddox, "Down with the Big Bang," *Nature* 340 (1989): 425.

[201] G. Burbidge, "Was There Really a Big Bang?" *Nature* 233 (1971): 36–40. G. F. R. Ellis, "Alternatives to the Big Bang," *Annual Review in Astronomy and Astrophysics* 32 (1984): 157–184. J. V. Narlikar, *The Primeval Universe* (Oxford University Press, Oxford, 1988), p. 225. H. C. Arp, G. Burbidge, F. Hoyle, J. V. Narlikar, N. C. Wickramasinghe, "The Extra-Galactic Universe: An Alternative View," *Nature* 346 (1990): 807–812. J. Maddox, "Big Bang Not Yet Dead but in Decline," *Nature* 377 (1995): 99. D. Berlinski, "Was There a Big Bang?" *Commentary* 105 (February 1998): 28–38.

[202] C. Seelig, *Helle Zeit—Dunkle Zeit*, p. 72.

serts a *creatio continua*, a continuous creation that can be interpreted as a divine interaction that goes on perpetually in order to sustain the expanding universe.

John Polkinghorne, a physicist and Anglican priest, did indeed maintain that "theology could have lived with either physical theory, for the assertion that God is Creator is not a statement that at a particular time he did something, but rather that at all times he keeps the world in being."[203]

The question of whether the generally accepted version of the Big Bang theory supports the doctrine of a divine creation of the universe has recently become the subject of debates not only among physicists and theologians but among philosophers as well. A noteworthy example is the debate between the philosophers Adolf Grünbaum and William Lane Craig. Grünbaum, who is well known for his monumental *Philosophical Problems of Space and Time* (1962, 1973), insists that the problem of the *origin* of the universe at a definite moment of time, say at $t_o = 0$, should not be confused with the problem of the creation of the universe *ex nihilo* at that time by an external cause. Although the former is a genuine problem, the latter is merely a pseudo-problem. Grünbaum first considered cosmological models that treat the Big Bang as a process that had been preceded by an infinite sequence of prior contractions and expansions and concluded that such theories of an eternally oscillating universe do "not even provide a point of departure for the argument from creation ex nihilo."[204] However,

[203] J. Polkinghorne, *Science and Creation* (Shambhala, Boston, 1998), p. 54.

[204] A. Grünbaum, "The Pseudo-problem of Creation in Physical Cosmology," *Philosophy of Science* 56 (1989): 373–394, reprinted in J. Leslie, ed., *Physical Cosmology and Philosophy* (Macmillan, New York,

the majority of modern cosmologists regard the Big Bang as a one-time event at $t_o = 0$ or immediately thereafter. Grünbaum distinguished two classes of such theories. Theories of the first class feature a cosmic time interval that is *closed* at the Big Bang instant t_o (i.e., $t_o \leq t$) and in which t_o of course had no temporal predecessor.

> This means that there simply did not exist any instant of time before $t_o = 0$! But it would be misleading to describe this state of affairs by saying that "time began" at $t_o = 0$. This description makes it sound as if time began in the same sense in which, say, a musical concert began. And that is misleading, precisely because the concert was actually preceded by actual instants of time, when it had not yet begun.[205]

Grünbaum could have mentioned that Immanuel Kant had already used essentially the same argument, namely that "the beginning is an existence which is preceded by a time in which the thing is not" in his discussion of the first antinomy when he argues in the antithesis that "the world has no beginning."[206] Clearly, any question like "what happened *before* $t = 0$?" or "what *caused* the Big Bang to occur

1990), pp. 92–112; "Pseudo-creation of the Big Bang," *Nature* 344 (1990): 821–822. "Creation as Pseudo-explanation in Current Physical Cosmology," *Erkenntnis* 35 (1991): 233–254.

[205] A. Grünbaum, "The Pseudo-problem of Creation in Physical Cosmology," p. 389.

[206] "Die Welt hat keinen Anfang . . . Denn man setze: sie habe einen Anfang. Da der Anfang ein Dasein ist, wovor eine Zeit vergeht, darin das Ding nicht ist, so muss eine Zeit vorhergegangen sein, darin die Welt nicht war." *Kritik der Reinen Vernunft*, 2d ed. (1787), in *Kant's Werke* (Akademie Ausgabe, G. Reimer, Berlin, 1911), vol. 3, p. 295. *Critique of Pure Reason* (Doubleday, Garden City, N.Y., 1961), p. 263.

at $t = 0$?" denies the physical correctness of the cosmological model.

In theories of the second class, the cosmic time interval is *open* at the Big Bang instant t_o (i.e., $t_o < t$) so that there is no first instant of time, or any instant of time prior to all instants of time in the cosmic time interval. Obviously, the illegitimacy of all those questions like "What caused the Big Bang?" holds in this case a fortiori and it can be said that the universe has always existed. In the sequel to his article, Grünbaum also referred to Einstein's general theory with respect to the inflational expansion in quantum gravitational models, such as Hawking's, and concluded "that neither the big bang cosmogony nor the steady-state cosmology validates the traditional cosmological argument for divine creation."

Craig challenged Grünbaum's analysis of the concept of "beginning"—and, we may add, Kant's—on the grounds that the beginning of something, say *x*, at a definite moment of time does not necessarily entail the existence of temporal instants before that moment of time. Referring to Grünbaum's example of a concert, Craig declared, "Imagine that the temporal instants prior to a performance of Beethoven's Fifth Symphony were nonexistent. Should we say that the symphony concert then fails to have a beginning?" To explain his objection, Craig argued that in Grünbaum's definition of "x begins to exist" his definition "*x* exists at time *t* and there are times immediately prior to *t* at which *x* does not exist" is wrong and should be replaced by the definition "*x* exists at *t* and there is no time immediately prior to *t* at which *x* exists."[207] Although Craig

[207] W. L. Craig, "The Origin and Creation of the Universe: A Reply

agreed to distinguish between "origin" and "creation" and was aware that his proposed definition of "x begins to exist" may need some further modification for, if applied to "God," it would entail the unwanted statement that "God begins to exist." He concluded that the question of the creation of the universe is a genuine problem that deserves discussion.[208]

In a rejoinder to Craig's critique, Grünbaum explained that, in accordance with the event-ontology of the general theory of relativity, the pathological topological properties of the Big Bang deprive it from having the status of an event in that theory because such a status requires chrono-geometric relations within a well-defined space-time metric. Hence, any model, at least of the first class, is not a bona fide cosmological model within the framework of general relativity and "the past physical career of the Big Bang did not include a first physical event or state at which it could be said to have begun."[209] Because only events can qualify as effects of other events, the Big Bang singularity "cannot be the effect of any cause in the case of either event-causation or agent-causation alike . . . and cannot have a cause, either earlier or simultaneous, be it natural or supernatural!"

---

to Adolf Grünbaum," *British Journal of the Philosophy of Science* 43 (1992): 233–240; "Prof. Grünbaum on Creation," *Erkenntnis* 40 (1994): 325–341.

[208] Craig's formulation of Grünbaum's definition of "*x* begins to exist" would face an even more serious difficulty if applied to the notion of "time." For the statement of the Big Bang cosmology that time begins, say at t_0, would mean, if x denotes "time," that "time exists at t_0 and there are times immediately prior to t_0 at which time does not exist"!

[209] A. Grünbaum, "Some Comments on William Craig's 'Creation and Big Bang Cosmology,'" *Philosophia Naturalis* 31 (1994): 226–236.

Grünbaum also challenged the argument for an external or supernatural creation based on the idea that the bare existence and persistence of the world must have a *ratio essendi* or that, as Craig contended, "the cause of the origin of the universe is causally prior to the Big Bang, though not temporally prior to the Big Bang." Grünbaum rejected such arguments on the ground that an asymmetric simultaneous causation is never encountered in nature.

Needless to say, in spite of their ideological opposition, Craig and Grünbaum highly respected each other. The same can be said about the sustained debate that has been fought out in an extended series of articles between Craig and the philosopher Quentin Smith. In fact, in 1993 they jointly coauthored a 350-page book containing a representative selection of their most important essays on the problem of whether the Big Bang cosmology, based on the Friedman solution of Einstein's field equations, warrants belief in a divine creation of the universe.[210]

Although almost every one of the issues dealt with in the Craig-Smith debate is somehow related to Einstein's theoretical work, only a few of them are discussed here. For example, Craig based his argument that "the universe has a cause of its existence" on the contention that "an infinite temporal regress of events cannot exist," and that contention, in turn, on the premise that "an actual infinite cannot exist." Finally, to substantiate this premise, he argued that the mathematical theory of sets, as developed by Cantor, Zermelo, Fraenkel, and others, "carries with it no ontological import for the existence of an actual infinite in

[210] W. L. Craig and Q. Smith, *Theism, Atheism, and Big Bang Cosmology* (Clarendon Press, Oxford, 1993).

the real world."[211] Smith challenged the theistic interpretation of the Big Bang cosmology not only on the grounds that it has a viable competitor in a nontheistic interpretation, but, more importantly, because, in his view, it is inconsistent with this cosmology. His argument, in brief, runs as follows. According to the Big Bang cosmology, there exists an earliest state, *E*, of the universe, which, by Hawking's so-called principle of ignorance, does not guarantee to evolve into an animate state. In the theistic interpretation, *E* is created by an omniscient, omnipotent, and perfectly benevolent God and must therefore evolve into an animate state; for an animate universe is better than an inanimate one. Hence, if the universe would not evolve into an animate state, God would be limited in his benevolence. Such a necessary evolution is incompatible with Hawking's principle.[212]

In a later essay, not reprinted in the coauthored book, Smith again invoked God's goodness, by virtue of which He would not deceive human beings—an idea reminiscent of René Descartes's *Méditations* (Meditation V, 8)—in order to disprove the theistic contention that God actually created the universe. There is abundant observational evidence, Smith pointed out, that confirms Hawking's theory that the universe began to exist without an external cause. "If God nevertheless caused the universe to begin to exist, he would have created a universe to come into existence in

[211] W. L. Craig, "The Finitude of the Past and the Existence of God," ibid., pp. 3–76. First published in W. L. Craig, *The Kalam Cosmological Argument* (Macmillan, London, 1979).

[212] Q. Smith, "Atheism, Theism, and Big Bang Cosmology," ibid., pp. 195–217. First published in *Australasian Journal of Philosophy* 69 (1991): 48–65.

this way. This would constitute a deception of human beings and would defeat the quintessentially human project of rational scientific inquiry. . . . However, God is good . . . and thus would not deceive human beings."[213]

The words with which Hawking concluded the exposition of his quantum-gravitational cosmological theory—"What place, then for a creator?"—have often been interpreted as a denial of the existence of God. In his last contribution to the collaborative book with Smith, Craig argued that such an interpretation of Hawking's position is quite misleading and false.[214] "For while it is true," says Craig, "that he [Hawking] rejects God's role as Creator of the universe in the sense of an efficient cause producing an absolutely first temporal effect, nevertheless Hawking appears to retain God's role as the Sufficient Reason for the existence of the universe, the final answer to the question, Why is there something rather than nothing?"[215]

According to Craig, Hawking failed to recognize that there are two kinds of creation, *creatio originans* and *creatio continuans*, the former being the act of bringing reality into being at a moment of time before which no such reality existed, the latter denoting God's preservation of this reality from moment to moment. Craig argued that even if Hawking's cosmological model were successful in eliminating the *creatio originans*, it would still need the *creatio continuans*, be-

[213] Q. Smith, "Stephen Hawking's Cosmology and Theism," *Analysis* 54 (1994): 236–243.

[214] W. L. Craig and Q. Smith, *Theism, Atheism, and the Big Bang Cosmology*.

[215] W. L. Craig, "'What Place, Then, for a Creator?': Hawking on God and Creation," in W. L. Craig and Q. Smith, *Theism, Atheism, and Big Bang Cosmology*, pp. 279–300. First published in *British Journal for the Philosophy of Science* 41 (1990): 473–491.

cause, without the continual bestowal of existence, "the whole of finite reality would be instantly annihilated and lapse into non-being." Hawking did not deny the existence of a Supreme Being, as can be seen from the following episode. Having "thoroughly enjoyed" reading an article by Hawking, published in the *American Scientist*,[216] James J. Tanner wrote a Letter to the Editors of this periodical stating that he was puzzled by the fact that "scientists are afraid to admit the existence of a Supreme Being whenever the subject of the universe arises." "Dr. Hawking pointed out," continued Tanner, "that 'all the evidence we have suggests the universe evolves in a well-determined way according to certain laws.'"[217] Hawking replied:

> I thought I had left the question of the existence of a Supreme Being completely open in my article. It would be perfectly consistent with all we know to say that there was a Being who was responsible for the laws of physics. However, I think it could be misleading to call such a Being 'God,' because this term is normally understood to have personal connotations which are not present in the laws of physics.[218]

If we recall the basic tenets of Einstein's cosmic religion and, in particular, his essay "Science and Religion," in which he expressed his belief in an impersonal Supreme Being, we should not be surprised if, in writing these lines, Hawking was influenced by Einstein's philosophy of religion.

[216] S. W. Hawking, "The Edge of Space-Time," *American Scientist* 72 (1984): 355–359.

[217] J. J. Tanner, "Letter to the Editors," *American Scientist* 73 (1985): 12.

[218] S. W. Hawking, ibid.

Einstein, it will be recalled, was interested most deeply in the question of "whether God could have made the world different; or in other words, whether the requirement of logical simplicity allows at all any choice" [ob die Forderung der logischen Einfachheit überhaupt eine Freiheit lässt]. Hawking raised a similar, though slightly more general, question when he wrote: "The usual approach of science of constructing a mathematical model cannot answer the questions of why there should be a universe for the model to describe. Why does the universe go to all the bother of existing?"[219] In other words, "Why is there something rather than nothing?" And Hawking added, "If we find the answer to that, it would be the ultimate triumph of human reason—for then we should know the mind of God."[220] But can science ever provide an answer to that question? It may be that the answer to this question belongs to a "region" that lies beyond the limits of physical science and its mathematical equations, just as Einstein said with respect to singularities, "All we have to realize is that the equations may not be continued over such regions."[221]

[219] S. W. Hawking, *A Brief History of Time*, p. 174.
[220] Ibid., p. 175.
[221] A. Einstein, "On the 'Cosmological Problem,'" p. 123.

Appendix

Zu Spinozas Ethik

Wie lieb ich diesen edlen Mann
Mehr als ich mit Worten sagen kann.
Doch fuercht' ich, dass er bleibt allein
Mit seinem strahlenden Heiligenschein.

So einen armen kleinen Wicht
Den fuehrst Du zu der Freiheit nicht.
Der amor dei laesst ihn kalt
Das Leben zieht ihn mit Gewalt.

Die Hoehe bringt ihm nichts als Frost
Vernunft ist fuer ihn schale Kost.
Besitz und Weib und Ehr' und Haus
Das fuellt ihn von oben bis unten aus.

Du musst schon guetig mir verzeihn
Wenn hier mir faellt Muenchhausen ein,
Dem als Einzigen das Kunststueck gediehn
Sich am eigenen Zopf aus dem Sumpf zu zieh'n.

Du denkst sein Beispiel zeigt uns eben
Was diese Lehre den Menschen kann geben.
Vertraue nicht dem troestlichen Schein:
Zum Erhabenen muss man geboren sein.

Einstein Archive, reel 33-264.

Index